Optical Nanomanipulation
(Second Edition)

Optical Nanomanipulation (Second Edition)

David L Andrews and David S Bradshaw
School of Chemistry, University of East Anglia, Norwich, Norfolk, UK

IOP Publishing, Bristol, UK

ISBN 978-0-7503-4191-2 (ebook)
ISBN 978-0-7503-4189-9 (print)
ISBN 978-0-7503-4192-9 (myPrint)
ISBN 978-0-7503-4190-5 (mobi)

DOI 10.1088/978-0-7503-4191-2

Version: 20220301

IOP ebooks

British Library Cataloguing-in-Publication Data: A catalogue record for this book is available from the British Library.

Published by IOP Publishing, wholly owned by The Institute of Physics, London

IOP Publishing, Temple Circus, Temple Way, Bristol, BS1 6HG, UK

US Office: IOP Publishing, Inc., 190 North Independence Mall West, Suite 601, Philadelphia, PA 19106, USA

To my wonderful daughter Elspeth and her husband Jack

—David L Andrews

To my nana in a difficult year

—David S Bradshaw

Contents

Preface from first edition

Among the numerous phenomena that were discovered within the first decade of the laser, one of the most striking was the pioneering demonstration by Arthur Ashkin, working at Bell Laboratories, that a laser beam could controllably manipulate microparticles. It was a proof of principle that paved the way for the technological breakthrough that became known as optical tweezers, which opened up a whole range of research applications; it also led to the development of atom cooling and subsequent trapping techniques. In fact, the discovery proved to be a tipping point for the identification of a remarkably extensive variety of phenomena in which laser light can be utilised to manipulate micro- and nano-sized particles. The whole field has grown and developed almost beyond recognition over the last fifty years, and it now represents a cutting edge of modern optics.

Today, outside of the routine application of optical tweezers as a technique for cell sorting procedures in medical laboratories, there is an increasing trend towards using light to locate, manoeuvre, rotate and assemble a wide range of nanoparticles, microparticles and even large molecules. Meanwhile, at the research end of the spectrum, the availability of a much wider range of laser power and beam structures now facilitates a variety of other methods for handling individual atoms and small molecules. The non-contact nature of optical manipulation offers huge advantages for the investigation and fabrication of matter at nanoscale dimensions, and it enables many techniques that would be simply impossible using any other methodology.

This book provides a broad introductory survey of this remarkable field, aiming to establish and clearly differentiate its physical principles, and also to provide a snapshot portrait of many of the most prominent current applications. Primary emphasis is placed on developing an understanding of the fundamental photonic origin behind the mechanism that operates in each type of effect. To this end, the first few chapters introduce and develop core theory, focusing on the physical significance and source of the most salient parameters, and revealing the detailed interplay between the key material and optical properties. Where appropriate, both classical and photonic (quantum mechanical) representations are discussed. The number of equations is purposely kept to a minimum, and only a broad background in optical physics is assumed.

With copious examples and illustrations, each of the subsequent chapters then sets out to explain and exhibit the main features and uses of the various distinct types of mechanism that can be involved in optical nanomanipulation, including some of the very latest developments. To complete the scene, we also briefly discuss applications to larger, biological particles. Overall, this book aims to deliver to the non-specialist an amenable introduction to the technically more advanced literature on individual manipulation methods. Full references to the original research papers are given throughout, and an up-to-date bibliography is provided for each chapter, which directs the reader to other selected, more specialised sources.

We freely acknowledge that the studies we describe represent only a fraction of the field as a whole; the subject has quickly outgrown the scale that any review could conceivably capture in its entirety. It is a great pleasure to acknowledge the privilege of help, in our endeavours, from many of the top researchers in this topic. We are especially indebted to numerous individuals who have supplied us with original figures, most of which have not been published previously. In addition, we gratefully thank colleagues, especially Mathew Williams, for numerous helpful comments. We are greatly indebted to all. The responsibility for any errors is, of course, our own.

David L Andrews
David S Bradshaw
Norwich, October 2016

Preface to second edition

Since the publication of our first edition, interest in the area of optical manipulation has greatly expanded. To a significant degree, and entirely fittingly, this may be attributed to the belated award of the 2018 Nobel Prize in Physics to Arthur Ashkin. With the announcement of this award, which as such post-dated our first edition, Ashkin became the oldest Nobel Prize laureate (at that time) at the age of 96. At the beginning of 2021, we gladly accepted the invitation by the Institute of Physics Publishing to update our book. This request was also timely in another way; over the last five years, an increasingly diverse volume of research has been published in this field. Aiming to retain an easy-to-understand approach, the primary objective of the second edition is to expand upon the original content, continuing to present the state of the art to the non-specialist reader.

Outside of the new material we have added, which itself extends the length of the book by around 40 pages—principally in chapters 4, 8, 11 and 12—the opportunity has been taken to make a few adjustments to the overall flow of contents. Since a great deal of the new research is bio-optical in nature, one of the chapters is now solely dedicated to the 'Biological applications of optical forces' (chapter 7). Some of the plasmon-related material on structured light has also been relocated, and we have now included some especially striking illustrations, showing the diversity in the field of optical manipulation, in a reworked chapter 1. In total, there are 11 new figures. Additionally, minor text changes have been made throughout the book and the occasional small error has been corrected.

We are again indebted to numerous individuals: Estela Martín Badosa, Kishan Dholakia, Sonja Franke-Arnold, David Grier, Miles Padgett, Grover Swartzlander and Mario Montes Usategui. All are leading researchers in the field, who for the first and second editions have contributed original figures. In addition, we are especially thankful to Halina Rubinsztein-Dunlop and ColdQuanta for both providing a new, unpublished image for the second edition. Last but not least, we are grateful to our colleague Kayn Forbes for casting a fresh pair of expert eyes over the final copy.

David L Andrews
David S Bradshaw
Norwich, December 2021

Author biographies

David L Andrews

David L Andrews, Professor of Chemical Physics at the University of East Anglia, UEA, has for 40 years headed a group internationally recognized for developments in the quantum theory of optical interactions, photonics, nonlinear optics, and chiral interactions. In work on optical binding his group first mapped the structures of optically conferred energy landscapes. David has also developed fundamental theory for electronic energy transfer and optical vortex interactions. He is highly active in international scientific societies and was the 2021 President of SPIE—The International Society for Optics and Photonics. He is a Fellow of SPIE, the Institute of Physics, the Royal Society of Chemistry and Optica. His publications exceed 400 papers and over 20 books, recently including an *Introduction to Photon Science and Technology*, *Molecular Photophysics and Spectroscopy*, *The Angular Momentum of Light*, and *Structured Light for Optical Communication*.

David S Bradshaw

David S Bradshaw is an Honorary Researcher at the University of East Anglia, Norwich, in the UK. He graduated twice from the same university, first receiving a master's degree in chemical physics (which included a year at the University of Western Ontario, London, Canada) and then a PhD in theoretical chemical physics. Overall, David has co-written 87 research papers, all based on molecular quantum electrodynamics. In addition, he has authored five books including *Introduction to Photon Science and Technology*, *Quantum Computing Made Easy* and *68 Greats of Science: Hippocrates to Hawking*. He also acted as the Editor for the fifth volume ('Applications of Nanoscience') of the *Comprehensive Nanoscience and Nanotechnology*, second edition. His many long-running research interests include resonance energy transfer, optical binding, and nonlinear optics. David is a Member of the Institute of Physics.

List of symbols

$\langle\ \rangle_{\mathrm{T}}$	time average
$\langle\ \rangle_{\mathrm{R}}$	rotational average
$\mathbf{A}$	classical electromagnetic vector potential
$\hat{\mathbf{A}}$	electromagnetic vector potential operator
a	particle diameter
$\hat{a}$	photon annihilation operator
$\hat{a}^{\dagger}$	photon creation operator
$\mathbf{B}$	magnetic field vector
$\hat{\mathbf{B}}$	magnetic field operator
$\mathbf{b}$	magnetic polarisation unit vector
c	speed of light
c_d	drag coefficient
d	particle diffusion distance
$\mathbf{E}$	electric field vector
$\hat{\mathbf{E}}$	electric field operator
E_0	energy of initial particle state
E_I	energy of initial system state
E_r	energy of virtual particle state
$\mathbf{e}$	electric polarisation unit vector
$\|F\rangle$	final system state
$f_{l,p}$	normalised radial function
g	acceleration due to gravity
H_{rad}	classical radiation Hamiltonian
$\hat{H}$	Hamiltonian operator of total system
$\hat{H}_0$	Hamiltonian operator of unperturbed system
$\hat{H}_{\mathrm{int}}$	interaction Hamiltonian operator
$\hat{H}_{\mathrm{mat}}$	matter Hamiltonian operator
$\hat{H}_{\mathrm{rad}}$	radiation Hamiltonian operator
h	Planck constant
$\hbar$	reduced Planck constant
$\|I\rangle$	initial system state
I	irradiance (intensity) of light beam
I_{ω}	irradiance per unit frequency
$\mathbf{J}_{\mathrm{rad}}$	classical total angular momentum of radiation
$\hat{\mathbf{J}}_{\mathrm{rad}}$	total angular momentum operator of radiation
$\mathbf{k}$	wave-vector
k	magnitude of the wave-vector
k_{B}	Boltzmann constant
k_x	force constant for trap stiffness
L_{Abra}	orbital angular momentum of light through a medium (Abraham)
L_{Mink}	orbital angular momentum of light through a medium (Minkowski)
$\mathbf{L}_{\mathrm{rad}}$	classical orbital angular momentum of radiation
$\hat{\mathbf{L}}_{\mathrm{rad}}$	orbital angular momentum operator of radiation
L	left-handed (circularly polarised light)
l	topological charge of vortex beam
ℓ	Fourier component index
M_{FI}	matrix element

m	particle or molecular mass
$\hat{N}$	number operator
N'	probability density for trapped particle position
n_a	atom number density
n_ω	refractive index
n_ϕ	number of absorbed photons
$\mathbf{P}_{\text{rad}}$	classical linear momentum of radiation
$\hat{\mathbf{P}}_{\text{rad}}$	linear momentum operator of radiation
P	radiation pressure
p	radial index
p_a	mean atomic momentum
R	reflectivity of the substance
^{R}	right-handed (circularly polarised light)
$\|R\rangle$	intermediate system state
$\boldsymbol{r}$	position vector
r	off-axis radial distance
$\|r\rangle$	virtual state of particle
r_b	radius of beam ring
$\mathbf{S}_{\text{rad}}$	classical spin angular momentum of radiation
$\hat{\mathbf{S}}_{\text{rad}}$	spin angular momentum operator of radiation
$\mathbf{s}$	inter-particle displacement
s	separation distance between particles
$\|s\rangle$	excited state of two-level particle
T	absolute temperature
T_{c}	critical temperature
T_{min}	minimum temperature
$\hat{T}_0$	resolvent operator
t	time
tr( )	trace of a matrix
$\mathbf{u_k}$	unit vector of $\mathbf{k}$
$\mathbf{u_{kz}}$	unit vector of $\mathbf{k_z}$
$\|u\rangle$	particle excited state
V	arbitrary volume
$\mathbf{v}$	velocity
W	beam power
w_0	focused beam waist
z	axial position
α	scalar polarisability
α'	volume polarisability
α_ℓ	normal mode expansion variable
α_{ij}	transition polarisability tensor
Γ	process rate
γ	natural linewidth
ΔE	potential energy
∇	gradient operator (differential with respect to $\mathbf{r}$)
$\delta\mathbf{v}$	recoil velocity
ε_0	permittivity of free space
η	viscosity of medium
$^{\eta}$	polarisation state
λ	wavelength of light

$\hat{\boldsymbol{\mu}}$	electric dipole moment operator
$\boldsymbol{\mu}^{0u}$	transition dipole moment (between ground and excited state)
ρ	density of particle
ρ_0	density of medium
ρ_F	energy density of states
σ_x	variance of a Gaussian distribution curve
$\boldsymbol{\tau}$	optical torque vector
Φ	electromagnetic scalar potential
ϕ	azimuthal angle
Ω_s	half-width at half-maximum of a Lorentzian lineshape
ω	angular optical frequency

IOP Publishing

Optical Nanomanipulation (Second Edition)

David L Andrews and David S Bradshaw

Chapter 1

Nanomanipulation: why optical methods are best

In ancient times, forces of a non-contact and invisible kind were thought to be somehow magical. The most familiar and readily observed of these forces, those of a magnetic or electric nature, only succumbed to explanation through progressively scientific studies from the Middle Ages onwards. On the eventual discovery that light was, in fact, electromagnetic radiation, it was verified in 1901 that light itself can manipulate neutral particles. This effect is usually so miniscule that a viable use for it was at first presumed near impossible. However, after the arrival of the laser in the 1960s, the extremely high light intensities of such devices enabled development of the first application of light-induced forces, in the technique now known as optical tweezers. Since the early days, enormous research progress has been made. While the trapping of microparticles has become common, the size of particles that can be manipulated optically now spans a wide range of nanoscale dimensions. This chapter outlines and illustrates the issues that arise over different scales.

1.1 Non-contact forces

In everyday experience, movement in the world that we perceive is most often a result of contact forces. The slightest puff of breeze stirs the treetops only through direct physical contact between the leaves and the moving substance of the air. Such is the familiarity of the forces that we most commonly observe and deploy, that any occurrences of objects being physically moved without material contact seem intrinsically more exotic, and often rather enigmatic. Gravity, whose effects provide the most obvious example of producing motion without contact, was long considered a mystery; it succumbed to Newton's analysis without yielding much mechanistic insight, and it is indeed still frustratingly difficult to weave into a satisfactory Grand Unified Theory. At least for the present, gravity also remains a force beyond our capacity to control.

doi:10.1088/978-0-7503-4191-2ch1

Over many centuries—at least since the first use of a lodestone—forces of electric and magnetic origin have held a particular fascination. They manifest a non-contact nature, yet they are amenable to control and application. Such forces succumbed to explanation, through progressively scientific studies, as far back as the Middle Ages onwards. The eventual dawning of the knowledge, in the 19th century, that light is itself electromagnetic in nature provided a logical basis for the understanding that it can exert a mechanical force. Such a prospect is consistent with our relatively recent discoveries—that, at the atomic level, even what are known as 'contact' forces are also fundamentally electrodynamic in origin. This manifests in bulk behaviour via the effect of 'hidden' interactions between the elementary charges that comprise every piece of matter.

To discover demonstrable manifestations of light-induced forces nonetheless proved much more difficult than for electric or magnetic effects. Maxwell and Bartoli [1, 2] among others, showed how and why light should have a capacity to convey and confer momentum, but the experimental confirmation awaited studies by Lebedev [3] and by Nichols and Hull [4] at the turn of the twentieth century. These works confirmed the presence of such effects, but they also verified that conventional sources would produce radiation forces of so little magnitude as to be scarcely amenable to measurement—let alone of meaningful practical use—and such effects would most often be completely masked by thermal forces, and the commonly associated diffusion. Indeed, it was only with the arrival of intense light sources in the form of lasers, in the early 1960s, that the determination and exercise of control on optical forces first became a practicable proposition.

The rapid pace of development of laser systems quickly provided experimental access to previously unattainable levels of intensity, substantially as a result of the high degree of angular confinement and controllable direction in the source emission. Exploiting this feature, Ashkin's Nobel Prize winning experiments first revealed the remarkable possibilities that laser beams could controllably manipulate small, neutral particles of matter [5]. In this inaugural work, the problem of competing thermal forces was overcome by using relatively transparent particles (i.e. latex spheres) suspended in a relatively transparent medium (water). These spheres were observed to be simultaneously drawn to the beam axis, and also accelerated in the direction of the light: optical manipulation was demonstrated for the first time. It was a demonstration that heralded the advent and development of a host of new techniques, at a pace closely following the rapid growth of laser technology.

1.2 Issues of scale

In conventional optics, the minimum physical beamwidth that can be achieved by focusing light is generally prescribed by the Abbé diffraction limit [6]. This limit is of the order of a wavelength—and therefore usually somewhere in the nanometre (submicron) regime for optical frequencies. (A variety of means have indeed been found to overcome this limit in super-resolution microscopy [7–10], notably the Nobel Prize winning technique of stimulated emission depletion, STED, developed by Hell *et al* [11, 12].) Consequently, the use of focused laser beams offers scope to exert controllable, non-contact forces on micro- or nanometre-scale particles, including for example biological cells. Such non-contact forces are especially useful in the nano regime because, on this scale, physical contact with manipulative equipment will often introduce problems due to adhesion, surface friction, stiction and contamination etc.

As we consider particles of progressively smaller size, gravity—whose effect essentially scales down with the cube of particle diameter—soon ceases to represent the most significant obstacle. To completely offset its effect, micron sized particles can usually be stably suspended in a liquid—and this is indeed a common configuration for optical manipulation experiments. Here, viscous drag is the only major limitation to particle motion. As a remarkable illustration of effects in such media, directly due to the propagation of light, a cambered refractive object can experience an optical lifting force analogous to aerodynamic lift when placed within a uniform stream of light (figures 1.1 and 1.2)—though the basis for this action is refraction and reflection, rather than the Bernoulli principle that applies to aerofoils [13]. Other light controllable particles, which are even more intricate in design, are known as micromotors. An example of the latter, which can be fabricated using two-photon polymerization [14] is shown in figure 1.3. Micromotors can be used in optical manipulation techniques that exploit photo-activated materials [15–21]; the propulsion on these and larger scale nanoobjects can be augmented by photochemical processes [16, 17]. Indeed, by ultimately photochemical means, light can in fact separate microscopic particles or surfaces that are in direct physical contact, but the light intensities required are much higher: laser ablation operates on this principle [22]. Such effects, however, do not represent truly controllable particle manipulation at the nanoscale and, thus, their detail lies outside the scope of this book.

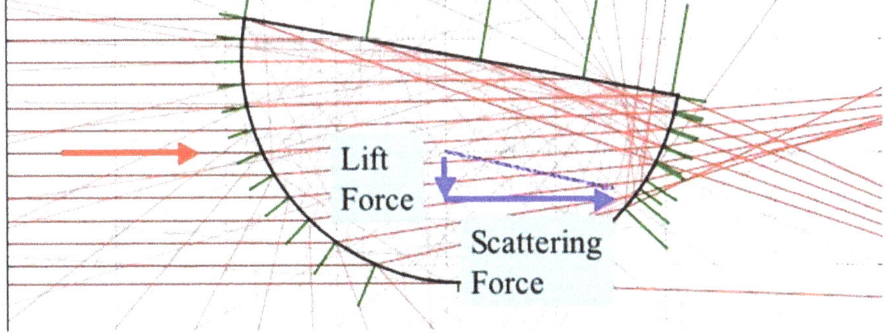

Figure 1.1. Cross-section of a uniformly illuminated semi-cylindrical rod, with the lift force and scattering (radiation) force indicated. Collimated rays (red) are incident from the left and the green lines represent the radiation force for each ray. Image courtesy of Grover Swartzlander, Rochester Institute of Technology.

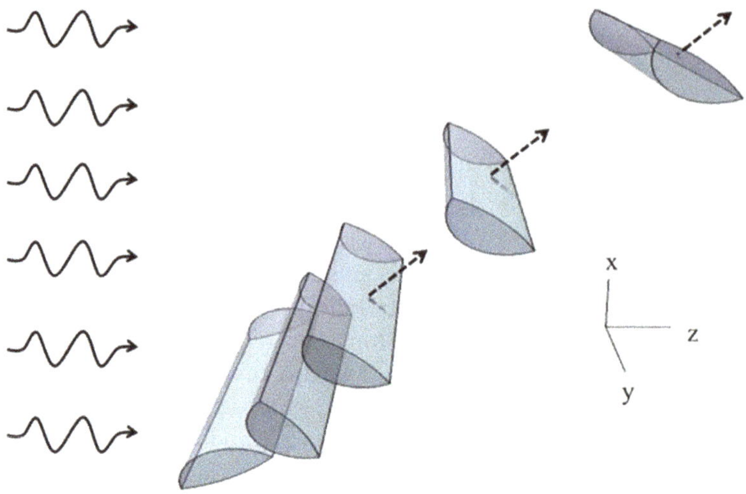

Figure 1.2. Artist's impression of the optical lift of a 'lightfoil' due to laser irradiation. Image courtesy of Grover Swartzlander, Rochester Institute of Technology.

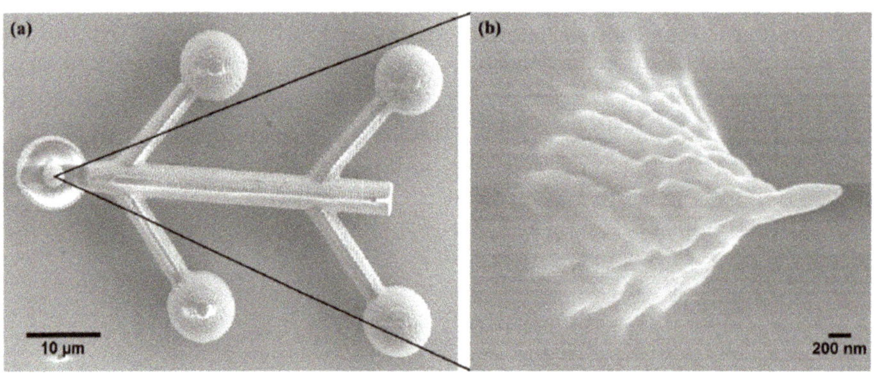

Figure 1.3. (a) Scanning electron microscopy image of a disk-tipped microrobot fabricated using two-photon polymerisation. Inset (b) shows a zoom-in of the tip. Reprinted with permission from reference [14]. Copyright 2018 SPIE.

There are other problems to confront as the dimensional scale is further reduced to atomic and molecule-sized particles. In gas and liquid phases, Brownian motion comes into play—powerfully so at ambient temperatures—serving as a reminder that at nanoscale level all matter is in a state of continual flux. Even at temperatures approaching absolute zero there is a residual energy of motion, reflecting the intrinsic quantum mechanical zero-point energy of all matter. The thermal energy that registers temperature is associated with the random kinetic motions of atoms and molecules. In gases at ambient temperatures, it is no longer viscosity but collisional effects that complicate the picture. For this reason, the optical manipulation of atoms and very small molecules is usually possible only at the extremely

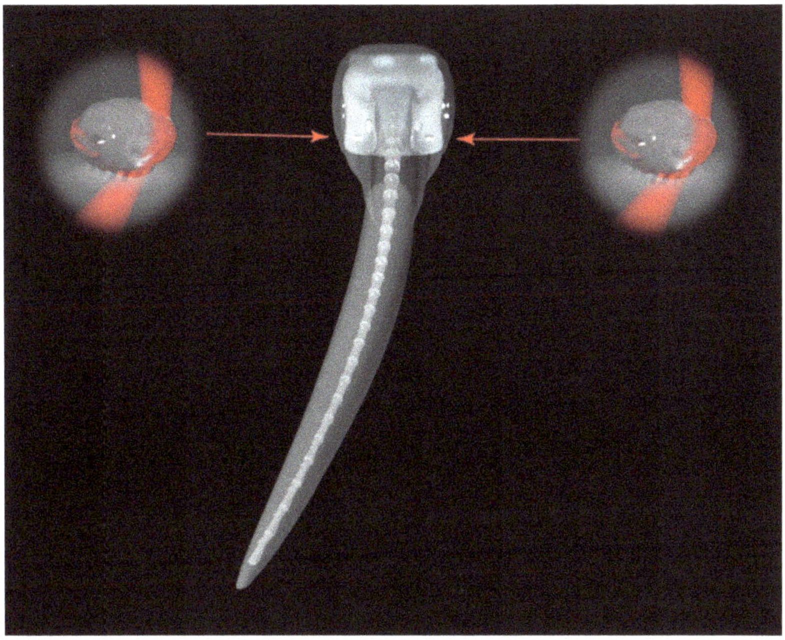

Figure 1.4. Directed laser light produces refraction in the otoliths of live zebrafish, creating controllable optical forces. The embryo responds to a perceived shift in balance by movement of its eyes and tail. Image courtesy of Halina Rubinsztein-Dunlop, University of Queensland.

low temperatures of ultra-cold gases (rather than liquid or solid states, since deposition or condensation would frustrate the sought particle mobility).

Due to the difficulties in scaling up ultra-cold methods, or scaling down microparticle trapping techniques, particles of intermediate scale (including quantum dots, nanowires and nanotubes) are optically manipulated using different approaches. So, as we shall see, both the methodologies and types of optical force used for nanomanipulation exist in numerous different forms, developed over the long history of the subject. Recent studies on live fish embryos spectacularly illustrate how far this field has evolved. Figure 1.4 depicts an ingenious scheme in which the neural processing of balance information is studied by non-contact optical manipulation of subcutaneous otolith particles, producing responsive movements of the creature's eyes and tail [23]: more detail is given in chapter 7. Numerous other spectacularly diverse applications are described in the pages that follow.

References

[1] Bartoli A 1884 Il calorico raggiante e il secondo principio di termodinamica *Il Nuovo Cimento* **15** 193–202

[2] Maxwell J C 1954 *A Treatise on Electricity and Magnetism* vol. 2 (New York: Dover Publications)

[3] Lebedev P N 1901 Experimental examination of light pressure *Ann. Phys.* (Berlin) **6** 433

[4] Nichols E F and Hull G 1901 A preliminary communication on the pressure of heat and light radiation *Phys. Rev.* **13** 307–20

[5] Ashkin A 1970 Acceleration and trapping of particles by radiation pressure *Phys. Rev. Lett.* **24** 156–9

[6] Abbe E 1873 Beiträge zur theorie des mikroskops *u*nd der mikroskopischen wahrnehmung *Arch. f. Mikro. Anat.* **9** 413–8

[7] Hell S and Stelzer E H K 1992 Properties of a 4Pi confocal fluorescence microscope *J. Opt. Soc. Am. A* **9** 2159–66

[8] Dunn R C 1999 Near-field scanning optical microscopy *Chem. Rev.* **99** 2891–928

[9] Gustafsson M G L 2000 Surpassing the lateral resolution limit by a factor of two using structured illumination microscopy *J. Microsc.* **198** 82–7

[10] Reymann J, Baddeley D, Gunkel M, Lemmer P, Stadter W, Jegou T, Rippe K, Cremer C and Birk U 2008 High-precision structural analysis of subnuclear complexes in fixed and live cells via spatially modulated illumination (SMI) microscopy *Chrom. Res.* **16** 367–82

[11] Hell S W and Wichmann J 1994 Breaking the diffraction resolution limit by stimulated emission: stimulated-emission-depletion fluorescence microscopy *Opt. Lett.* **19** 780–2

[12] Klar T A and Hell S W 1999 Subdiffraction resolution in far-field fluorescence microscopy *Opt. Lett.* **24** 954–6

[13] Swartzlander G A, Peterson T J, Artusio-Glimpse A B and Raisanen A D 2011 Stable optical lift *Nat. Photonics* **5** 48–51

[14] Bunea A-I, Engay E, Chouliara M, Bañas A R and Glückstad J 2018 Rational design of light-controlled microrobots *Proc. SPIE* **10804** 1080406

[15] Palima D and Glückstad J 2013 Gearing up for optical microrobotics: micromanipulation and actuation of synthetic microstructures by optical forces *Laser Photon. Rev.* **7** 478–94

[16] Eskandarloo H, Kierulf A and Abbaspourrad A 2017 Light-harvesting synthetic nano- and micromotors: a review *Nanoscale* **9** 12218–30

[17] Xu L, Mou F, Gong H, Luo M and Guan J 2017 Light-driven micro/nanomotors: from fundamentals to applications *Chem. Soc. Rev.* **46** 6905–26

[18] Wang J, Xiong Z, Zheng J, Zhan X and Tang J 2018 Light-driven micro/nanomotor for promising biomedical tools: principle, challenge, and prospect *Acc. Chem. Res.* **51** 1957–65

[19] Zhan Z, Wei F, Zheng J, Yang W, Luo J and Yao L 2018 Recent advances of light-driven micro/nanomotors: toward powerful thrust and precise control *Nanotech. Rev.* **7** 555–81

[20] Bunea A-I, Martella D, Nocentini S, Parmeggiani C, Taboryski R and Wiersma D S 2021 Light-powered microrobots: challenges and opportunities for hard and soft responsive microswimmers *Adv. Intell. Syst.* **3** 2000256

[21] Wang J, Xiong Z and Tang J 2021 The encoding of light-driven micro/nanorobots: from single to swarming systems *Adv. Intell. Syst.* **3** 2000170

[22] Shirk M D and Molian P A 1998 A review of ultrashort pulsed laser ablation of materials *J. Laser Appl.* **10** 18–28

[23] Favre-Bulle I A, Stilgoe A B, Rubinsztein-Dunlop H and Scott E K 2017 Optical trapping of otoliths drives vestibular behaviours in larval zebrafish *Nat. Commun.* **8** 630

David L Andrews and David S Bradshaw

Chapter 2

Key properties of the radiation

The key properties of light that relate to optical forces are its energy, linear momentum and angular momentum. Since optical manipulation over nanoscale dimensions clearly involves photon interactions, it makes sense to develop an understanding based on those self-same properties of the photon. These, in turn, depend on the fundamental electric and magnetic fields, whose quantisation leads to expressions in operator form. The propagation of light through materials modifies its wave-vector and so too its linear momentum, in ways that invite different interpretations (the Abraham and Minkowski representations). More generally, deriving a quantum amplitude for a given type of measurement secures a clear pathway for the calculation of optical forces of wide-ranging kinds.

2.1 Energy, linear momentum and angular momentum of light

Traditionally, the analysis of optical forces begins with a text-book derivation based on the electromagnetic *Poynting vector* [1] (as depicted for a plane wave in figure 2.1), working from its usual interpretation as the linear momentum density of a beam, cast in terms of the local electromagnetic fields. This connection enables beam properties such as intensity and polarisation to be linked to the effects exerted on matter in the course of beam obstruction, deflection etc. Notwithstanding recent deliberations over the precise status of this interpretation—an issue still open to discussion—we too shall establish this link later in this chapter. Nevertheless, as befits a focus on nanoscale interactions, it is appropriate to build upon a quantum foundation. This affords us a deeper and more fundamentally complete picture of light, well suited to addressing the mechanical interactions of electromagnetic fields with matter. To this end, we start by considering the properties of photons and establish the factors that are relevant for producing an optical force.

Each photon can be considered to convey—in the sense of quantifiable attributes ascribed to matter (an absorber, for example)—essentially three mechanical quantities: energy, linear momentum, and angular momentum. As a scalar, the radiation

doi:10.1088/978-0-7503-4191-2ch2

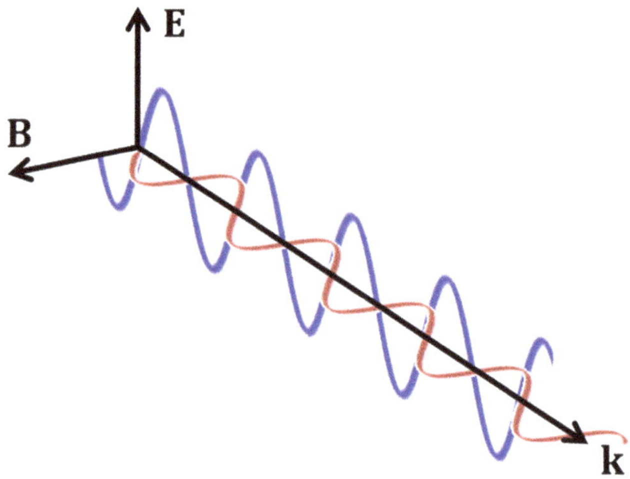

Figure 2.1. Linearly polarised electromagnetic radiation: the **E**, **B** and **k** orthogonal axes denote the directions of the electric field, magnetic field and propagation direction, respectively. The Poynting vector (which is proportional to $\mathbf{E} \times \mathbf{B}$) points in the **k** direction, the blue and red curves represent the displacement of the electric and magnetic fields, respectively.

energy lacks any intrinsic directionality. However, for a beam delivering a particular photon flux, the linear and angular momenta conferred in the process of photon absorption afford the primary basis for radiation pressure, as well as linear and angular forces (torques). Other kinds of force, generally secondary effects due to scattering, can be produced where the local potential energy landscape is shaped by a non-uniform beam intensity distribution.

A definition of each of these photon quantities, and the relationships between them, is now appropriate. Moreover, it proves helpful to introduce the classical beam representations alongside their quantum counterparts. The two are, of course, formally linked. Securing equations cast in full quantum form is made possible through a promotion of the classical expressions into operator form, from which the corresponding properties of individual photons emerge. For simplicity at this stage, the key equations are developed for free-space propagation; issues associated with local fields in the condensed phase are addressed later. At this juncture, readers who have less interest in the following, concise summary of the formal development may skip ahead to chapter 3.

We begin with *classical* expressions for the generalised position- and time-dependent electric and magnetic free fields, $\mathbf{E}(\mathbf{r},t)$ and $\mathbf{B}(\mathbf{r},t)$ respectively, associated with electromagnetic radiation. Developing a Fourier series representation, each field can be written as a sum of two mode-summed contributions [2];

$$\mathbf{E}(\mathbf{r},t) = \mathbf{E}^{(+)}(\mathbf{r},t) + \mathbf{E}^{(-)}(\mathbf{r},t); \quad \mathbf{B}(\mathbf{r},t) = \mathbf{B}^{(+)}(\mathbf{r},t) + \mathbf{B}^{(-)}(\mathbf{r},t), \tag{2.1}$$

whose constituent terms are referred to as the positive and negative frequency parts of the fields, respectively. The signs in the superscripts relate to spatial phase factors that feature in the explicit Fourier expansions: the terms $\mathbf{E}^{(+)}(\mathbf{r},t)$ (typically known as the *analytic signal*) and $\mathbf{E}^{(-)}(\mathbf{r},t)$ are given by;

$$\mathbf{E}^{(+)}(\mathbf{r},t) = \mathrm{i}\sum_{\ell} \mathbf{e}_{\ell} E_{\ell} \alpha_{\ell}(t) \mathrm{e}^{\mathrm{i}(\mathbf{k}_{\ell}\cdot\mathbf{r})};\ \mathbf{E}^{(-)}(\mathbf{r},t) = \mathrm{i}\sum_{\ell} \bar{\mathbf{e}}_{\ell} E_{\ell} \alpha_{\ell}^{*}(t) \mathrm{e}^{-\mathrm{i}(\mathbf{k}_{\ell}\cdot\mathbf{r})}, \tag{2.2}$$

where the Fourier component with index ℓ relates to a set of four numbers relating to the wave-vector $\mathbf{k}_{\ell}$ and polarisation unit vector $\mathbf{e}_{\ell}$, $\alpha_{\ell}(t)$ is a normal variable dependent on the positive form of the phase factor and E_{ℓ} is a mode constant. Expressions for the two components of $\mathbf{B}(\mathbf{r},t)$ have a similar form. To satisfy Maxwell's equations, the time-dependent factor $\alpha_{\ell}(t)$ has to carry a temporal phase factor $\exp(-\mathrm{i}\omega t)$; similarly $\alpha_{\ell}{}^{*}(t)$ carries a temporal phasor $\exp(\mathrm{i}\omega t)$. Thus, individual components of the Fourier series can be regarded as representing a beam of angular optical frequency $\omega = ck$, propagating with a wave-vector $\mathbf{k}$. For single-mode applications, the direction of $\mathbf{k}$ is usually taken to define the z-axis.

To determine analogous expressions in a quantum framework, i.e. where the electric and magnetic fields are quantised, the factor α_{ℓ} and its conjugate are promoted to the status of operators on radiation states—the operator form to be signified here, and henceforth, by the introduction of a caret (^) symbol—to obtain [3];

$$\begin{aligned}\hat{\mathbf{E}}(\mathbf{r}) &= \hat{\mathbf{E}}^{(+)}(\mathbf{r}) + \hat{\mathbf{E}}^{(-)}(\mathbf{r}) \\ &= \mathrm{i}\sum_{\mathbf{k},\eta}\left(\frac{\hbar c k}{2\varepsilon_0 V}\right)^{\frac{1}{2}}\left\{\mathbf{e}^{(\eta)}(\mathbf{k})\hat{a}^{(\eta)}(\mathbf{k})\mathrm{e}^{\mathrm{i}(\mathbf{k}\cdot\mathbf{r})} - \bar{\mathbf{e}}^{(\eta)}(\mathbf{k})\hat{a}^{\dagger(\eta)}(\mathbf{k})\mathrm{e}^{-\mathrm{i}(\mathbf{k}\cdot\mathbf{r})}\right\},\end{aligned} \tag{2.3}$$

$$\begin{aligned}\hat{\mathbf{B}}(\mathbf{r}) &= \hat{\mathbf{B}}^{(+)}(\mathbf{r}) + \hat{\mathbf{B}}^{(-)}(\mathbf{r}) \\ &= \mathrm{i}\sum_{\mathbf{k},\eta}\left(\frac{\hbar k}{2\varepsilon_0 c V}\right)^{\frac{1}{2}}\left\{\mathbf{b}^{(\eta)}(\mathbf{k})\hat{a}^{(\eta)}(\mathbf{k})\mathrm{e}^{\mathrm{i}(\mathbf{k}\cdot\mathbf{r})} - \bar{\mathbf{b}}^{(\eta)}(\mathbf{k})\hat{a}^{\dagger(\eta)}(\mathbf{k})\mathrm{e}^{-\mathrm{i}(\mathbf{k}\cdot\mathbf{r})}\right\},\end{aligned} \tag{2.4}$$

where the summation is taken over each photon with mode $(\mathbf{k}, \eta)$, in which η labels the polarisation state, c is the speed of light in a vacuum, ε_0 is the permittivity of free space, $\hbar$ is the reduced Planck constant and V is an arbitrary quantisation volume—a notional region within which the system of interest resides. The Hermitian conjugate pair of operators $\hat{a}^{(\eta)}(\mathbf{k})$ and $\hat{a}^{\dagger(\eta)}(\mathbf{k})$ signify photon annihilation and creation, respectively, i.e. they reduce or increase by one the number of photons present; the electric and magnetic polarisation vectors of each photon, denoted by $\mathbf{e}^{(\eta)}(\mathbf{k})$ and $\mathbf{b}^{(\eta)}(\mathbf{k}) = \mathbf{e}^{(\eta)}(\mathbf{k}) \times \mathbf{k}$ (overbars signifying their complex conjugates), also appear in the above expressions.

At this juncture, it is noteworthy that the sum over polarisations can be effected over any pair that constitutes an orthogonal basis: this will usually be two orthogonal linear polarisations (often designated 'vertical' and 'horizontal' with respect to a reference plane) or left- and right-handed circular polarisation. In fact, any two polarisation states that correspond to diametrically opposite positions on a Poincaré sphere may form an acceptable basis. In the quantum formalism, $\hat{\mathbf{E}}(\mathbf{r})$ and $\hat{\mathbf{B}}(\mathbf{r})$ are not dependent on time since, in the assumed Schrödinger representation, any time dependence is held within the state vector.

The quantum energy operator is developed, in the usual way, from an expression for the classical Hamiltonian of the radiation, H_{rad}. The latter entails an integral of the energy density over volume, namely;

$$H_{\text{rad}} = \frac{\varepsilon_0}{2}\int_V (|\,\mathbf{E}(\mathbf{r},t)\,|^2 + c^2\,|\,\mathbf{B}(\mathbf{r},t)\,|^2)\mathrm{d}^3r. \tag{2.5}$$

On substitution of the quantum field operators, i.e. equations (2.3) and (2.4), into this expression—where V again acquires the significance of a quantisation volume—the following Hamiltonian operator may be found;

$$\hat{H}_{\text{rad}} = \sum_{\mathbf{k},\eta}\left(\hat{N}^{(\eta)}(\mathbf{k}) + \frac{1}{2}\right)\hbar\omega, \tag{2.6}$$

where the dependence of ω on $\mathbf{k}$ is left implicit; $\hat{N}^{(\eta)}(\mathbf{k}) = \hat{a}^{\dagger(\eta)}(\mathbf{k})\hat{a}^{(\eta)}(\mathbf{k})$ is the *number operator*, which returns the number of photons when applied to a number (Fock) state. Moreover, for the ground state of the radiation, $\frac{1}{2}\hbar\omega$ represents the zero-point energy, associated with vacuum fluctuations that are present even in the absence of radiation [4]. The latter feature, a well-known characteristic of the quantum formalism, proves to have very significant physical connotations—as we shall see in a later chapter. By inspection of equation (2.6), it can be determined that the energy of a single photon is $\hbar\omega$.

The classical linear momentum of the radiation, $\mathbf{P}_{\text{rad}}$, also involves an integral of the Poynting vector over V:

$$\mathbf{P}_{\text{rad}} = \varepsilon_0\int_V (\mathbf{E}(\mathbf{r},t) \times \mathbf{B}(\mathbf{r},t))\mathrm{d}^3\mathbf{r}\,. \tag{2.7}$$

Here we take the opportunity to emphasise that, in the simple mode expansions represented above, the Poynting vector is directed along the propagation axis. This is true for *plane waves*, whose wavefront comprise regularly spaced planar surfaces separated by a wavelength. This assumption signifies a *paraxial approximation*, applicable to beams with negligible angular dispersion—being neither significantly convergent nor divergent. It is an approximation that works well for most kinds of nano-optical interaction. However, with structured or strongly focused beams, where the wavefront is curved (or may even be helicoidal, as we shall see in chapter 9), optical manipulation often exploits features resulting from the fact that Poynting vectors in different transverse regions of the beam depart from exact collinearity.

To continue: the total angular momentum of the radiation, $\mathbf{J}_{\text{rad}}$, is proportional to an integral of the cross product between the Poynting vector and displacement, i.e.;

$$\mathbf{J}_{\text{rad}} = \varepsilon_0\int_V \mathbf{r} \times (\mathbf{E}(\mathbf{r},t) \times \mathbf{B}(\mathbf{r},t))\mathrm{d}^3\mathbf{r}\,. \tag{2.8}$$

For laser beams, the above expression can, in turn, be re-written as the sum of two terms, given explicitly as the respective terms in the following;

$$\mathbf{J}_{\text{rad}} = \varepsilon_0\int_V (\mathbf{E}(\mathbf{r},t) \times \mathbf{A}(\mathbf{r},t) + E_i(\mathbf{r},t)(\mathbf{r} \times \nabla)A_i(\mathbf{r},t))\mathrm{d}^3\mathbf{r} = \mathbf{S}_{\text{rad}} + \mathbf{L}_{\text{rad}}. \tag{2.9}$$

Here, $\mathbf{A}(\mathbf{r},t)$ is the electromagnetic vector potential, and the initial and final terms of the equation represent the orbital and spin angular momentum, respectively. (In principle, the form of separation given by equation (2.9) is not unique, since the electromagnetic vector potential is not gauge-invariant. In the Coulomb gauge used here, the separation is strictly applicable only on the assumption that optical beams have paraxial form: for strongly focused beams the two separate parts of equation (2.9) do not support exact quantization, as spin–orbit coupling occurs [5].) In the latter term, the repeated index i represents summation (also known as an *Einstein summation* [6]) over any frame of Cartesian coordinates.

The quantum operator form of the former, spin term of equation (2.9) involves the following mode expansion of the vector potential:

$$\hat{\mathbf{A}}(\mathbf{r}) = \sum_{\mathbf{k},\eta}\left(\frac{\hbar}{2\varepsilon_0 ckV}\right)^{\frac{1}{2}}\left\{\mathbf{e}^{(\eta)}(\mathbf{k})\hat{a}^{(\eta)}(\mathbf{k})\mathrm{e}^{\mathrm{i}(\mathbf{k}\cdot\mathbf{r})} - \bar{\mathbf{e}}^{(\eta)}(\mathbf{k})\hat{a}^{\dagger}(\eta)(\mathbf{k})\mathrm{e}^{-\mathrm{i}(\mathbf{k}\cdot\mathbf{r})}\right\}. \tag{2.10}$$

This potential is related to the electric and magnetic fields via $\mathbf{E} = -\partial\mathbf{A}/\partial t - \nabla\Phi$ and $\mathbf{B} = \nabla\times\mathbf{A}$, respectively, where ∇ denotes a differential with respect to $\mathbf{r}$ and Φ is the electromagnetic scalar potential. Inserting the field operators (2.3), (2.4) and (2.10) into the corresponding classical expressions $\mathbf{P}_{\mathrm{rad}}$, $\mathbf{S}_{\mathrm{rad}}$ and $\mathbf{L}_{\mathrm{rad}}$ (as required) results in the production of their quantum equivalents. Namely, after a little manipulation, we find that;

$$\hat{\mathbf{P}}_{\mathrm{rad}} = \sum_{\mathbf{k},\eta}\hat{N}^{(\eta)}(\mathbf{k})\hbar\mathbf{k}, \tag{2.11}$$

in which the zero-point momentum for each mode, $\frac{1}{2}\hbar\mathbf{k}$, vanishes due to each $\mathbf{k}$ in the summation having a matching $-\mathbf{k}$ (c.f. the zero-point energy, which persists since energy is always a positive quantity). Accordingly, it is established that the linear momentum is $\hbar\mathbf{k}$ for any photon with wave-vector $\mathbf{k}$. It also emerges that radiation states of pure circular polarisation are eigenstates of the operator for quantum spin, $\mathbf{S}_{\mathrm{rad}}$. This is consistent with an angular momentum of $\pm\hbar$ per photon with the sign determined by handedness; the vector quantity directed along $\mathbf{k}$ is thus expressible as $\pm\hbar\mathbf{u}_{\mathrm{k}}$, where $\mathbf{u}_{\mathrm{k}}$ represents a unit vector of $\mathbf{k}$. This spin feature is concisely expressed by the following identity [4];

$$\hat{\mathbf{S}}_{\mathrm{rad}} = \sum_{\mathbf{k}}\left\{\hat{N}^{(L)}(\mathbf{k}) - \hat{N}^{(R)}(\mathbf{k})\right\}\hbar\mathbf{u}_{\mathrm{k}}, \tag{2.12}$$

with summation specifically cast in terms of circular polarisations, either left- or right-handed light as denoted by the superscript L and R. Equation (2.12) shows, as might be expected, that the spin-angular momentum operator, $\hat{\mathbf{S}}_{\mathrm{rad}}$, depends solely on a disparity in the number of left- and right-handed photon [7, 8]. The explicit result for $\hat{\mathbf{L}}_{\mathrm{rad}}$ is deferred until chapter 9, where the appropriate mode structures are introduced.

In summary, the quantum development of theory shows that in addition to an energy of $\hbar\omega$, each photon conveys a linear momentum $\hbar\mathbf{k}$ and, in the case of circular polarisations, a unit spin angular momentum of $+\hbar$ for left-handed and $-\hbar$ for right-handed light.

2.2 Light inside a medium

The previous section implicitly assumed that the light traverses a vacuum. To account for electromagnetic radiation propagating within a medium or condensed phase, it is expedient to develop the fields (in either a classical or quantum form) to include a Lorenz factor, $(n_\omega^2 + 2)/3$ [9, 10], where n_ω is the refractive index of the dielectric medium. In the quantum case, these factors are formally derived from first principles. The momentum of light in a dielectric material is nonetheless the source of a century-old dispute known as the Abraham–Minkowski controversy. The debate centres on the Minkowski prediction that an increase in optical linear momentum should arise on entry of an electromagnetic wave into a dielectric medium, while Abraham predicted a decrease: the underlying logic is encapsulated in figure 2.2. Depending on the observable that an experiment is designed to measure, results can be obtained that are consistent with each of the two possibilities [11, 12]. Meanwhile, for a considerable period of time, theoretical studies were unable to agree on the correct solution [13]. The controversy is now considered to some extent resolved. Since the momentum of a photon in a dielectric medium is an abstract concept that is experimentally unmeasurable, only the total momentum of the material and light combined is detectable—therefore, both predictions may be viewed as applicable. As a result, the division of total momentum into component (optical and material) parts is arbitrary and may be performed to produce a kinetic or canonical momentum [14–20].

The two corresponding formulations for orbital angular momentum are $L_{\text{Mink}} = l\hbar$ and $L_{\text{Abra}} = l\hbar/n_\omega^2$, where l is the topological charge (detailed in chapter 9) and the former equation is identical to the free-space form. The spin angular momentum of circularly polarised light will, in an isotropic medium, have the same dependence on n_ω. The correct formalism for angular momentum is yet to be fully resolved; Padgett *et al* have expressed support for the Abraham formulation [21]. A comprehensive analysis has recently been provided by Bliokh *et al* [22].

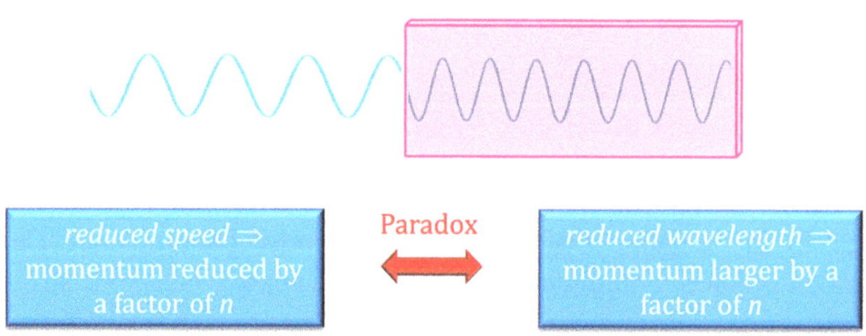

Figure 2.2. Illustration depicting the paradoxically opposite interpretations by Abraham (left-hand box) and Minkowski (right-hand box) of the momentum change in light entering a medium.

2.3 Matter and its interaction with light

So far, only the character of the electromagnetic radiation has been examined. Before describing the optical forces that light may exert on materials, the key properties of matter and its interaction with light are now scrutinised. Alongside obvious mechanical features—for instance, mass and moment of inertia—there are specifically electrodynamical properties, such as polarisability. The latter concerns the local electric response of the delocalised electron cloud of the material, which has a specifically quantum mechanical basis. Material properties of such a quantum origin are directly measurable; moreover, they are generally expressible by formulae that allow for their calculation, based on the fundamental principles of quantum mechanics. In anticipation of later detail, it is useful to recognise the typical treatment of such features.

Let us consider a system in which optical radiation engages with matter in some fashion. The usual starting point of a quantum analysis is to write down an expression for the Hamiltonian $\hat{H}$, i.e. the energy operator for the complete system. We recall that in simple quantum mechanics the 'system' usually comprises matter alone, and the state wavefunctions are secured as eigenstates of the corresponding $\hat{H}_{\text{mat}}$. When radiation enters the scene, the system Hamiltonian also contains a term $\hat{H}_{\text{rad}}$, which is the Hamiltonian for the radiation. However, whenever optical interactions occur there is a further additional term that needs consideration, namely the interaction Hamiltonian $\hat{H}_{\text{int}}$. Thus, the full quantum expression for the system Hamiltonian can be expressed as the sum of three parts;

$$\hat{H} = \hat{H}_{\text{mat}} + \hat{H}_{\text{rad}} + \hat{H}_{\text{int}}. \tag{2.13}$$

The last of these three terms is usually treated as a perturbation on the light and matter states. Theory, based on this perturbation, is then developed from the following general expression for the *matrix element*, M_{FI}, for progression from an initial system state $| I \rangle$ to a final state $| F \rangle$;

$$M_{FI} = \sum_{q=0}^{\infty} \langle F | \left\{ \hat{H}_{\text{int}} \left(\hat{T}_0 \hat{H}_{\text{int}} \right)^q \right\} | I \rangle. \tag{2.14}$$

In the above equation $\hat{T}_0$, known as a *resolvent operator*, is given by $\hat{T}_0 \approx (E_I - \hat{H}_0)^{-1}$, where the unperturbed system is represented by $\hat{H}_0 = \hat{H}_{\text{mat}} + \hat{H}_{\text{rad}}$ and E_I is the energy of the initial system state.

A flowchart that assists the identification of the physical observable of a given matrix element is presented in figure 2.3. The primary consideration is whether the quantum state of the material system differs from its initial state, as a result of an optical interaction. This possibility includes quantum transitions produced by absorption or inelastic scattering. A distinction is then drawn, in the case where no such material transition occurs, between optical processes that observably change the radiation state—such as deflective scattering—from those that do not. As shown, these criteria together determine how the matrix element is developed into an expression for a direct observable.

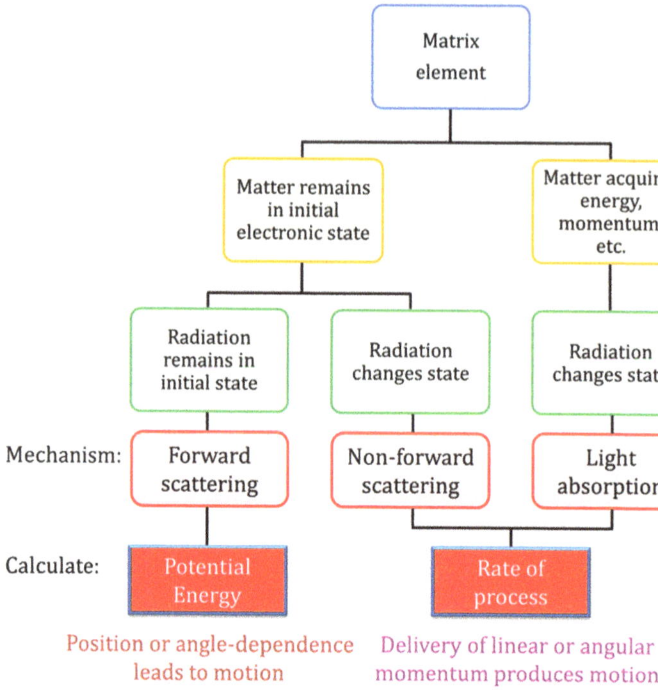

Figure 2.3. Flowchart exhibiting essential differences in implementation of the matrix element. The different pathways of interpretation are determined by considering any change in the quantum state of the matter and radiation; the relevant mechanism and its measurable are then deduced.

Following application of equation (2.14), all of the key electrodynamical properties can be derived, and also analytical expressions for the observables associated with optical processes—such as rates, intensities and forces. For example, the $q = 0$ term delivers transition dipole moments and oscillator strengths for absorption processes, while the $q = 1$ term gives an explicit expression for polarisability, as we shall see below [4]. In the latter connection, one clear value of these methods is the identification of the precise dependence on the frequency of the irradiating light. The importance of all these factors in the determination of optical forces will emerge in the following chapters.

References

[1] Poynting J H 1884 On the transfer of energy in the electromagnetic field *Philos. Trans. R. Soc. A* **175** 343–61

[2] Grynberg G, Aspect A and Fabre C 2010 *Introduction to Quantum Optics: From the Semi-Classical Approach to Quantized Light* (Cambridge: Cambridge University Press) p. 314

[3] Andrews D L and Bradshaw D S 2018 *Introduction to Photon Science and Technology* (Bellingham, WA: SPIE Press)

[4] Bradshaw D S and Andrews D L 2017 Manipulating particles with light: radiation and gradient forces *Eur. J. Phys.* **38** 034008

[5] Bliokh K Y, Dressel J and Nori F 2014 Conservation of the spin and orbital angular momenta in electromagnetism *New J. Phys.* **16** 093037

[6] Einstein A 1916 Die grundlage der allgemeinen relativitätstheorie *Ann. Phys.* (Berlin) **354** 769–822

[7] Andrews D L and Coles M M 2012 Measures of chirality and angular momentum in the electromagnetic field *Opt. Lett.* **37** 3009–11

[8] Coles M M and Andrews D L 2012 Chirality and angular momentum in optical radiation *Phys. Rev. A* **85** 063810

[9] Juzeliūnas G 1996 Microscopic theory of quantization of radiation in molecular dielectrics: normal-mode representation of operators for local and averaged (macroscopic) fields *Phys. Rev. A* **53** 3543–58

[10] Juzeliūnas G 1997 Microscopic theory of quantization of radiation in molecular dielectrics. II. Analysis of microscopic field operators *Phys. Rev. A* **55** 929–34

[11] Padgett M J 2008 On diffraction within a dielectric medium as an example of the Minkowski formulation of optical momentum *Opt. Express* **16** 20864–8

[12] Zhang L, She W, Peng N and Leonhardt U 2015 Experimental evidence for Abraham pressure of light *New J. Phys.* **17** 053035

[13] Pfeifer R N C, Nieminen T A, Heckenberg N R and Rubinsztein-Dunlop H 2007 Momentum of an electromagnetic wave in dielectric media *Rev. Mod. Phys.* **79** 1197–216

[14] Barnett S M 2010 Resolution of the Abraham–Minkowski dilemma *Phys. Rev. Lett.* **104** 070401

[15] Baxter C and Loudon R 2010 Radiation pressure and the photon momentum in dielectrics *J. Mod. Opt.* **57** 830–42

[16] Mansuripur M 2010 Resolution of the Abraham–Minkowski controversy *Opt. Commun.* **283** 1997–2005

[17] Milonni P W and Boyd R W 2010 Momentum of light in a dielectric medium *Adv. Opt. Photon.* **2** 519–53

[18] Kemp B A 2011 Resolution of the Abraham-Minkowski debate: implications for the electromagnetic wave theory of light in matter *J. Appl. Phys.* **109** 111101

[19] Ramos T, Rubilar G F and Obukhov Y N 2015 First principles approach to the Abraham–Minkowski controversy for the momentum of light in general linear non-dispersive media *J. Opt.* **17** 025611

[20] Brevik I 2018 Analysis of recent interpretations of the Abraham–Minkowski problem *Phys. Rev. A* **98** 043847

[21] Padgett M J, Barnett S M and Loudon R 2003 The angular momentum of light inside a dielectric *J. Mod. Opt.* **50** 1555–62

[22] Bliokh K Y, Bekshaev A Y and Nori F 2017 Optical momentum and angular momentum in complex media: from the Abraham–Minkowski debate to unusual properties of surface plasmon-polaritons *New J. Phys.* **19** 123014

IOP Publishing

Optical Nanomanipulation (Second Edition)

David L Andrews and David S Bradshaw

Chapter 3

Optically induced mechanical forces

There are two forms of distinct optical forces that act on neutral nanoparticles. One is traditionally associated with radiation pressure, which is the effect discovered by Maxwell and Bartoli. This force develops because momentum is imparted onto the material when light is either absorbed or re-directed by matter. The other arises as a Lorentz force, in response to a radiation-induced potential energy that is introduced wherever the light intensity varies with particle position. At the quantum level, this effect owes its origin to the production of intensity-dependent internal energy level shifts, associated with the dynamic Stark effect. The quantum mechanisms are interpreted, and their theory derived, using Feynman diagrams.

3.1 Overview

The electric and magnetic fields of light may generate optical forces due to their engagement with the charge distributions in suitably irradiated nanoparticles. The effect is most obvious under resonance conditions, where a photon energy matches the quantum change in electronic energy of its absorber: the resulting optical transition will generally result in a modification of the internal energy distribution, at the atomic or molecular level. However, electromagnetic fields can also exert a more subtle effect under non-resonance conditions. Here, as we shall see, their capacity to produce short-lived *virtual* transitions lies at the heart of other important kinds of optical force. In these cases, light–matter interactions result in only transient changes to the internal electronic energy of the material. Then, according to the location of the matter relative to the local distribution of the light, forces may result from the position-dependent energy. In terms of photonics, optical trapping and manipulation of neutral particles by a laser beam may, therefore, involve two completely independent mechanisms.

3.2 Radiation forces

One form of optical force, the subject of this section, is traditionally associated with radiation pressure. As established in the previous chapter, electromagnetic radiation

doi:10.1088/978-0-7503-4191-2ch3

has an intrinsic linear momentum determined from the Poynting vector. When the light is absorbed, or the direction of the radiation is altered by its interaction with matter, light momentum is imparted onto the material in the form of a *radiation force*. Perhaps misleadingly, the term *scattering force* is sometimes applied to both forms of radiation force.

For many years following the original discovery, no practical applications were considered feasible for the extremely small radiation forces associated with terrestrial-level solar or artificial light. Such effects were only considered necessary to account for the much larger solar radiation pressures observed in an astronomical context [1–6]. With the advent of continuous-wave lasers, however, with tightly collimated beams presenting much higher levels of intensity, it was soon found practicable for optical forces to be routinely used to accelerate, decelerate, deflect, guide and trap micro- and nano-particles [7]. It is useful to recognise the connection between the essentially classical, macroscopic and more obviously quantum level nanoscopic forces associated with optical radiation.

Classical description. On the macroscopic scale, assuming the irradiation of a smooth, flat and opaque material, a quantitative expression for what is now known as the Maxwell–Bartoli force can readily be derived [8, 9]. Since an optical force can be equally described by energy per unit distance, and momentum per unit time (from Newton's second law), it follows that optical momentum can be defined as energy divided by the speed of light. The radiation pressure, P, is the rate of change of momentum divided by the irradiated area so that: $P = I\cos^2\theta/c$, where I is the irradiance—the energy arriving at the surface per unit time (input power) per unit area, and θ is the angle between the surface normal and the incident radiation. This expression describes the radiation pressure acting on a perfectly absorbing surface, which is illustrated by figure 3.1(a). If, however, the surface is perfectly reflecting, then a factor of two is included on the right-hand side of the expression, since momentum is exerted by both the incident and reflected light, as shown by figure 3.1(b). In practice, the pressure delivered to any real surface will lie somewhere between these two extremes, according to its reflectivity R. Hence, the general Maxwell–Bartoli expression emerges as follows:

$$P = (1 + R)(I/c)\cos^2\theta. \tag{3.1}$$

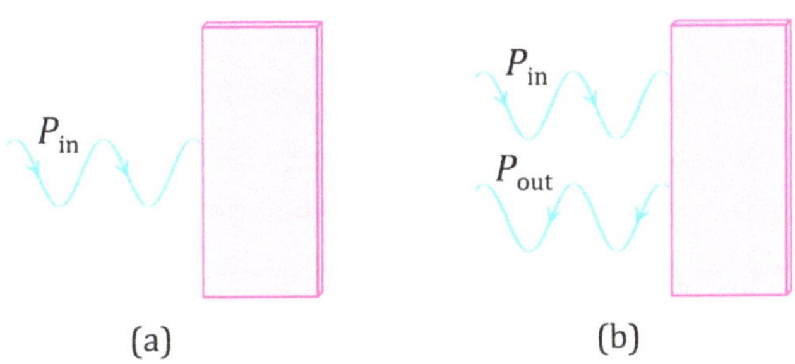

Figure 3.1. Radiation pressure, determined from the Maxwell–Bartoli expression, applied to a surface due to (a) absorption (R = 0) and (b) reflection (R = 1): P_{in} and P_{out} represent the radiation pressure due to the incident and reflected light, respectively.

Quantum description. At the quantum level, the radiation force when a photon is absorbed by a surface particle can be scaled up for the entire surface by multiplying the rate of photon absorption, Γ, by the linear momentum $n_{\phi}\hbar\mathbf{k}$, where n_{ϕ} denotes the number of absorbed photons. The rate of absorption, for a given wave-vector $\mathbf{k}$ and polarisation η, is found from the Fermi rule, $\Gamma = (2\pi\rho_F/\hbar)|M_{FI}|^2$, where ρ_F is the energy density of states and M_{FI} is the matrix element determined from first-order time-dependent perturbation theory [10]. The result emerges from the $q = 0$ term in equation (2.14) as:

$$M_{FI} = \langle F \mid \hat{H}_{\text{int}} \mid I \rangle = \langle (n-1); u \mid -\hat{\boldsymbol{\mu}} \cdot \hat{\mathbf{E}}(\mathbf{r}) \mid n; 0 \rangle. \tag{3.2}$$

Here, using Dirac bracket notation, $| I\rangle = | n; 0\rangle$ and $| F\rangle = | (n-1); u\rangle$ represent the specific initial and final system states, respectively, decomposed into radiation and particle components: the initial state contains n photons and the particle is in the ground state 0, and the final state (which follows photon absorption) comprises n–1 photons and the particle in its excited state u. The given number of photons in each case represents the quantity within the notional quantisation volume V; only their ratio, which relates to irradiance, will feature in the emerging results for any observable. At this juncture, it is appropriate to introduce the Feynman diagrammatic representation; the diagram corresponding to equation (3.2) is shown in figure 3.2.

The structure of equation (3.2) reveals that it is a change in state of the radiation field that allows for the transfer of momentum from the irradiating light to the nanoparticle. Also appearing in equation (3.2) is the explicit expression for the interaction Hamiltonian, i.e. $\hat{H}_{\text{int}} = -\hat{\boldsymbol{\mu}}\cdot\hat{\mathbf{E}}\,(\mathbf{r})$ where $\hat{\boldsymbol{\mu}}$ is the electric dipole moment operator that acts on the particle states to produce a transition dipole moment $\boldsymbol{\mu}^{u0}$ and $\hat{\mathbf{E}}(\mathbf{r})$ acts on the radiation states. Here, in fact, only the $\hat{\mathbf{E}}^{(+)}(\mathbf{r})$ portion of equation (2.3)

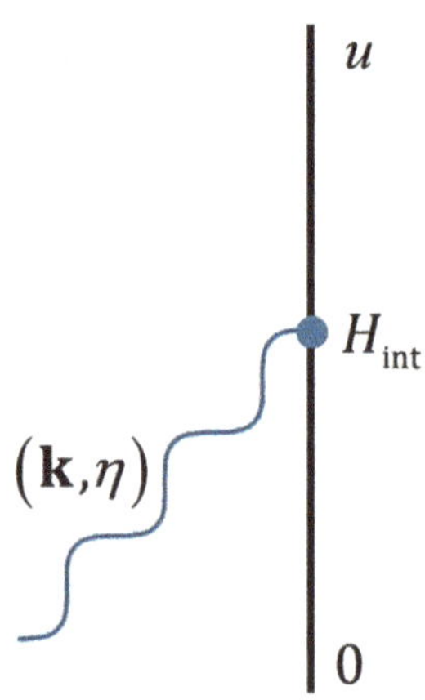

Figure 3.2. Feynman diagram illustrating the radiation force due to single-photon absorption. The blue wavy line represents a photon of mode ($\mathbf{k}$, η), the vertical line symbolises the electronic state of a particle that begins in a ground state 0 and finishes in its excited state α, the blue dot denotes the light–matter interaction, labelled by H_{int}, and time travels upwards. Since the initial state $| I\rangle = | n; 0\rangle$ and final state $| F\rangle = | (n-1); u\rangle$ differ, this diagram represents a process in which the physical observable is a rate, delivered from the matrix element using the Fermi rule.

is required, since it is this term that includes the photon annihilation operator. Assuming $\mathbf{e}\|\boldsymbol{\mu}^{u0}$ and with the density of states given by $\rho_F = I_\omega V/2\pi cn\hbar^2\omega$ [11], the radiation force is determined as [12];

$$\mathbf{F} = \frac{n_\phi I_\omega \mathbf{k}}{2c\varepsilon_0\hbar} \mid \boldsymbol{\mu}^{u0} \mid^2, \tag{3.3}$$

where I_ω is the irradiance per unit frequency. In cases where the incident photons are deflected, i.e. scattered rather than absorbed, the treatment of theory is a little more intricate and involves second-order perturbation theory; this inelastic process is presented in chapter 9.

Photon absorption may also impart *angular* momentum, primarily (for conventional laser light) according to its state of polarisation, as we saw in the previous chapter. However, such angular momentum is transferred to the internal electronic motions associated with the change in electronic state. The well-known bulk transfer of optical angular momentum from circularly polarised light leading to rotation, as in Beth's classic experiment [13] or more recent equivalents [14, 15], is a transmission (i.e. forward-scattering) effect.

The other elementary optical process that can, in principle, impart momentum at nanoscale dimensions is the time-inverse of photon absorption, i.e. emission. However, when emission is a spontaneous phenomenon in free space, the reactive force that acts on the emitter does not produce any overall effect due to the random directions in which such emission occurs. This principle was indeed exploited in early experiments on laser isotope separation [16], the selected isotope being preferentially deflected out of an atomic beam through multiple absorption and emission processes. We shall examine this technique, in its use for atomic gas cooling, more closely in the next chapter.

3.3 Gradient force

Another form of optical mechanism that results in the centre-of-mass motion of a free particle is commonly known as the *gradient force* (or, alternatively, the *dipole force*). At the atomic or molecular level this involves the production of intensity-dependent internal energy level shifts, associated with the dynamic (ac) Stark effect. In quantum mechanical terms, this arises from the local perturbation of electronic wavefunctions by the electric field of the radiation, through an interaction in which the latter features quadratically. In such a case, since the level shifts depend on the intensity of the light, an effective potential energy surface is introduced wherever that intensity varies with the particle position—which is usually across the cross-section of a beam and, thus, it is related to the transverse field distribution. The gradient force is determined from the spatial derivative of the potential energy.

Classical description. In terms of classical electrodynamics, the gradient force is commonly defined in terms of the Lorentz force on a point dipole which, introducing a constant of proportionality (a scalar polarisability) α, is expressed as [17, 18];

$$\mathbf{F}(\mathbf{r},t) = \alpha\left(\frac{1}{2}\nabla E(\mathbf{r},t)^2 + \frac{\mathrm{d}}{\mathrm{d}t}(\mathbf{E}(\mathbf{r},t) \times \mathbf{B}(\mathbf{r},t))\right). \tag{3.4}$$

The second term relates to the Poynting vector—which can again be cast, using equation (2.1), as a sum of four terms involving $\mathbf{E}^{(+)}(\mathbf{r},t)$ and $\mathbf{E}^{(-)}(\mathbf{r},t)$ in cross-products with their magnetic counterparts. It then transpires that the product contributions $\mathbf{E}^{(+)}(\mathbf{r},t)\mathbf{B}^{(-)}(\mathbf{r},t)$ and its complex conjugate are explicitly time-independent, and hence persistent: the two other terms carry factors $\exp(\pm 2i\omega t)$, and hence oscillate too rapidly to engage with whole-particle motion. The 'Poynting' term in equation (3.4) thus disappears, and the time-averaged optical force (denoted below by $\langle\ \rangle_T$) becomes [19];

$$\langle \mathbf{F}\,(\mathbf{r})\rangle = \frac{\alpha}{2}\nabla\langle E(\mathbf{r})^2\rangle_{\mathrm{T}}. \tag{3.5}$$

Here, the electric field is involved in a quadratic response and, therefore, we can conclude that the gradient force is sustained in an oscillating field (static fields often relate to a linear response).

Quantum description. Turning again to the quantum theory, it is apparent that in the gradient force (unlike the radiation force) the electromagnetic field suffers no change in system state, i.e. no transfer of linear momentum arises. Here, the position-dependent potential energy, ΔE, is evaluated as an expectation value, in which the initial and final system states are identical, so that the interaction involves photon annihilation and recreation into the same radiation mode—this mechanism is known as forward-Rayleigh (elastic) scattering, as depicted in figure 3.3, wherein

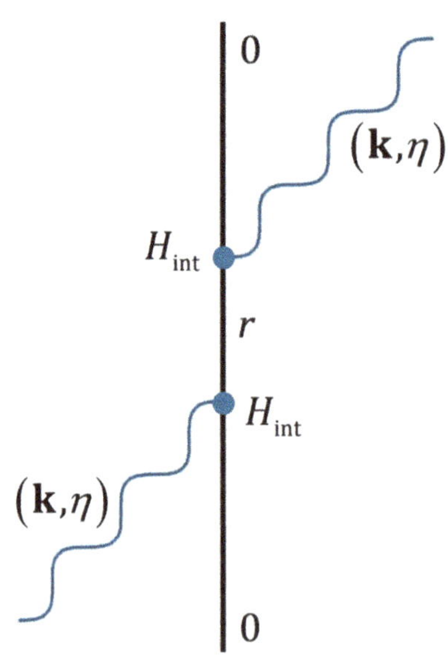

Figure 3.3. Feynman diagram illustrating the gradient force based on forward-Rayleigh scattering, in which a photon is annihilated and another recreated into the same radiation mode. The particle begins and ends in the ground state, traversing through a short-lived virtual state r. Since the initial and final state are identical, i.e. $|\,I\rangle = |\,F\rangle = |\,n;\,0\rangle$, this diagram represents a potential energy obtained from the real part of the matrix element.

the frequency of the trapping beam is non-resonant with any energy level transitions in the nanoparticle.

Interpreting figure 3.3, using the framework of second-order perturbation theory, i.e. the $q = 1$ term in equation (2.14), the following matrix element signifies the evaluation of an expectation value;

$$\Delta E = \mathrm{Re}\, M_{II} = \mathrm{Re}\left\{\sum_R \frac{\langle I \mid \hat{H}_{\mathrm{int}} \mid R\rangle\langle R \mid \hat{H}_{\mathrm{int}} \mid I\rangle}{E_I - E_R}\right\}, \tag{3.6}$$

where $\mid I\rangle = \mid n;\, 0\rangle$ denotes both the initial and final system state and $\mid R\rangle = \mid n - 1;\, r\rangle$ is an intermediate system state; E is the energy of the state denoted by the subscript. It is important to note that $|R\rangle$ involves a short-lived virtual particle state r, rather than any specific excited state u. Moreover, $\hat{\mathbf{H}}_{\mathrm{int}} = -\hat{\boldsymbol{\mu}}\cdot\hat{\mathbf{E}}(\mathbf{r})$ will invoke either term of equation (2.3) so that either $\hat{\mathbf{E}}^{(+)}(\mathbf{r})$ or $\hat{\mathbf{E}}^{(-)}(\mathbf{r})$ appears in $\hat{H}_{\mathrm{int}}$, depending on whether the operator acts on a photon annihilation or creation event. Using a beam irradiance (in units of W m^{-2}) given by $I = n\hbar c^2 k/V$, the gradient force is found as [12];

$$\mathbf{F}(\mathbf{r}) = \left(\frac{\nabla I(\mathbf{r})}{2\varepsilon_0 c}\right)\alpha_{ij}(\omega)\bar{e}_i e_j. \tag{3.7}$$

Here, the η and $\mathbf{k}$ dependences are suppressed on the unit polarisation vector $\mathbf{e}$, and the polarisability $\alpha_{ij}(\omega)$ emerges, in a complete form, as a second rank property tensor explicitly given by;

$$\alpha_{ij}(\omega) = \sum_r \left\{\frac{\mu_i^{0r}\mu_j^{r0}}{E_{r0} - \hbar\omega} + \frac{\mu_j^{0r}\mu_i^{r0}}{E_{r0} + \hbar\omega}\right\}, \tag{3.8}$$

where $E_{r0} = E_r - E_0$. In passing, we note that results are more often cast in terms of the linear optical susceptibility, in place of the polarisability tensor, when light is scattered by larger nanoparticles.

Equation (3.7) is written in a concise form that uses the Einstein summation convention (for repeated tensor indices). In any application to three-dimensional space, this simply means that where subscript indices i, j are used to signify x, y or z components in a Cartesian frame of reference, the repetition of an index within a given term signifies an implicit summation over those same three directions. For example, a scalar product $\boldsymbol{\mu}\cdot\mathbf{E}$ may be written as $\mu_i E_i \equiv \mu_x E_x + \mu_y E_y + \mu_z E_z$ and the scalar $\alpha_{ii} \equiv \alpha_{xx} + \alpha_{yy} + \alpha_{zz} = \mathrm{tr}(\boldsymbol{\alpha})$. (To save any confusion, the rule also stipulates that no such index can appear more than twice in any given term.) It transpires that equation (3.7) tallies with equation (3.5) through a rotational average leading to $\left\langle \alpha_{ij}(\omega)\bar{e}_i e_j\right\rangle_{\mathrm{R}} = \frac{1}{3}\mathrm{tr}(\boldsymbol{\alpha}) = \alpha$ together with the relation $I = c\varepsilon_0|\mathbf{E}|^2$. Molecular polarisability is often reported as a 'volume polarisability' defined as $\alpha' = \alpha'\, 4\pi\varepsilon_0$ and, as such, the force per molecule that follows from equation (3.7) is of the order $2\alpha' I/c w_0$, where w_0 is the (focused) beam waist. For example $\alpha' = 10^{-29}$ m^3, $I = 10^{15}$ W m^{-2} and $w_0 = 1$ μm gives an optical force of around 0.1 fN, within measurable range.

The orientational effect of the gradient force is less well-known, and it is not evident if the target particles have spherical symmetry—nor can it be identified from the equations if a scalar representation of the polarisability is used. The majority of the micro- and nano-particles studied in optical traps are indeed of essentially spherical form, meaning that their polarisability tensor can be represented by a diagonal matrix with all three components identical. Nonetheless, there is interest in non-spherical particles, and the simplest case arises with particles of cylindrical or ellipsoidal symmetry. Here, the diagonal polarisability matrix has one value, $\alpha_{\|}$, for the axial direction, and another, $\alpha_{\perp}$, for each of the other two axes. It is noteworthy that $\alpha_{\|}$ will always have the larger magnitude in a prolate ellipsoid since, in the longer axial direction, the polarisability—which is, basically, a measure of the shift in electron distribution to engender a dipole moment—will correspond to a larger electron shift. (The reader should avoid possible confusion associated with the term 'polarisability ellipsoid', whose perverse historic definition is a surface with an inverse square root dependence on polarisability values in each direction.) Any such non-spherical particle, initially sitting at an arbitrary angle with respect to the linearly polarised input radiation, will experience a torque that tends to align its long axis with the polarisation plane, at right-angles to the direction of light propagation. An expression for the torque at the nanoscale is given by;

$$\boldsymbol{\tau} = (\alpha_{\perp} - \alpha_{\|})\left(E_y\mathbf{i} - E_x\mathbf{j}\right)E_z, \tag{3.9}$$

where E_x, E_y and E_z are the electric field directed in the x-, y- and z-directions, z being the long axis of the particle; $\mathbf{i}$ and $\mathbf{j}$ signify x- and y-unit vectors, respectively. For linearly polarised light, the torque becomes zero when there is complete alignment with the polarisation vector, such that $E_x = E_y = 0$. For any particle offset in the (x, z) plane the result is a torque about the y-axis, towards full alignment; a (y, z) offset exerts a restoring twist about the x-axis. An experimental example is shown in figure 5.3 in chapter 5. However, it is noteworthy that larger microscale particles with cylindrical symmetry will align with the propagation direction, in a beam of narrow waist, to maximise engagement with the high intensity region [20].

There is another less widely recognised mechanical effect of light, known as *electrostriction*, which occurs within the bulk of a solid through which light passes. It too depends on the electric field of the radiation engaging with the charges of which the material is comprised. As illustrated in figure 3.4, this effect can cause a small shrinkage in one direction and expansion in another, according to the orientation of its electric field. Its internal effect within nanoscale particles is too small to be significant, but the underlying mechanism is closely connected to *optical binding*, which relates to interactions *between* nanoparticles. This is an increasingly important phenomenon in the field of optical nanomanipulation; it is covered in chapter 12.

To summarise in terms of a quantum framework, the crucial difference in the calculation of the radiation and gradient force is that the former uses the rate as the physical observable, since the initial and final system states differ, while the latter is found from the potential energy in which the initial and final states are identical [21]—as

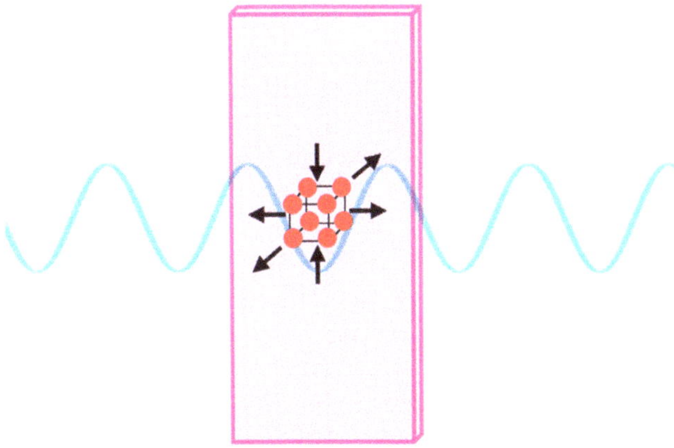

Figure 3.4. Electrostriction produced by light travelling through a solid. The constituent atoms of the solid are compressed together in the direction of the electric field (up and down arrows) and expanded in the other perpendicular directions (left and right arrows, in- and out-of-plane arrows).

indicated by the flowchart of figure 2.3. The radiation force involves the transfer of momentum from the beam to the nanoparticle through either light absorption or non-forward-Rayleigh scattering (where the incident and emergent photon have the same energy but travel in different directions). The gradient force implies an attraction of the nanoparticle to the high intensity part of the beam via forward-Rayleigh scattering (all photons here have the same energy and direction).

References

[1] Poynting J H 1904 Radiation in the solar system: its effect on temperature and its pressure on small bodies *Philos. Trans. R. Soc. A* **202** 525–52

[2] Robertson H P 1937 Dynamical effects of radiation in the solar system *Mon. Not. R. Astron. Soc.* **97** 423–37

[3] Ives N E 1963 The effect of solar radiation pressure on the attitude control of an artificial earth satellite *Aeronautical Research Council Reports & Memoranda* No. 3332

[4] Tsander F A 1964 *Problems of Flight by Jet Propulsion: Interplanetary Flights* (Jerusalem: Israel Program for Scientific Translations Ltd)

[5] Burns J A, Lamy P L and Soter S 1979 Radiation forces on small particles in the solar system *Icarus* **40** 1–48

[6] McInnes C R 2004 *Solar Sailing: Technology, Dynamics and Mission Applications* (Berlin: Springer)

[7] Ashkin A 1997 Optical trapping and manipulation of neutral particles using lasers *Proc. Natl Acad. Sci. USA* **94** 4853–60

[8] Bartoli A 1884 Il calorico raggiante e il secondo principio di termodinamica *Il Nuovo Cimento* **15** 193–202

[9] Maxwell J C 1954 *A Treatise on Electricity and Magnetism* vol. 2 (New York: Dover Publications)

[10] Cohen-Tannoudji C, Dupont-Roc J and Grynberg G 1992 *Atom–Photon Interactions: Basic Processes and Applications* (New York: Wiley) p. 59

[11] Craig D P and Thirunamachandran T 1998 *Molecular Quantum Electrodynamics: An Introduction to Radiation–Molecule Interactions* (Mineola, NY: Dover Publications) p. 89
[12] Bradshaw D S and Andrews D L 2017 Manipulating particles with light: radiation and gradient forces *Eur. J. Phys.* **38** 034008
[13] Beth R A 1936 Mechanical detection and measurement of the angular momentum of light *Phys. Rev.* **50** 115–25
[14] Moothoo D N, Arlt J, Conroy R S, Akerboom F, Voit A and Dholakia K 2001 Beth's experiment using optical tweezers *Am. J. Phys.* **69** 271–6
[15] Nieminen T A, Rubinsztein-Dunlop H and Heckenberg N R 2001 Calculation and optical measurement of laser trapping forces on non-spherical particles *J. Quant. Spectrosc. Radiat. Transfer* **70** 627–37
[16] Bernhardt A F 1976 Isotope separation by laser deflection of an atomic beam *Appl. Phys.* **9** 19–34
[17] Gordon J P 1973 Radiation forces and momenta in dielectric media *Phys. Rev. A* **8** 14–21
[18] Nieto-Vesperinas M, Chaumet P C and Rahmani A 2004 Near-field photonic forces *Philos. Trans. R. Soc. A* **362** 719–37
[19] Chaumet P C and Nieto-Vesperinas M 2000 Time-averaged total force on a dipolar sphere in an electromagnetic field *Opt. Lett.* **25** 1065–7
[20] Simpson S H and Hanna S 2007 Optical trapping of spheroidal particles in Gaussian beams *J. Opt. Soc. Am. A* **24** 430–43
[21] Bradshaw D S and Andrews D L 2013 Interparticle interactions: energy potentials, energy transfer, and nanoscale mechanical motion in response to optical radiation *J. Phys. Chem. A* **117** 75–82

IOP Publishing

Optical Nanomanipulation (Second Edition)

David L Andrews and David S Bradshaw

Chapter 4

Laser deflection, cooling and trapping of atoms

In many respects freely moving atoms represent an ideal system for the study and application of optical forces. The simplest method achieves deflection from an atomic beam, by direct transfer of linear momentum. More developed techniques for the optical trapping of atoms are also based on radiation forces acting at low temperature and pressure conditions. Laser light can be suitably configured to achieve extreme cooling, decelerating atomic motions essentially to a standstill, and so enabling ultraprecise spectroscopic measurements to be performed. With some atomic samples, the state of matter known as Bose–Einstein condensates may be formed, which exhibit collective quantum modes. Laser cooling is also now achievable with a select number of small molecules.

4.1 Atomic beam deflection

To directly study the deflective effect of laser light on atoms requires atomic beam instrumentation. Deflection is produced by irradiation across the atom beam, as a direct result of photon momentum transfer. Isotopic selectivity in this set-up can be achieved by the use of laser wavelengths in resonance with an electronic transition in one specific isotope. To this end, the isotope shift of the transition needs to be much larger than its natural width, and also larger than the laser linewidth; proofs of principle became well-established in the 1980s [1].

The absorption of a photon of visible light imparts a momentum in the order of 10^{-27} kg m s^{-1}, signifying that the increment in transverse velocity produced by each photon may be no more than 1 cm s^{-1}. Since atoms in an atomic beam may have an axial velocity of around 500 m s^{-1}, each such deflection is only significant at the milliradian level. However, an atomic excited state lifetime is typically measured in nanoseconds, and so a series of repeated excitation and decay cycles can occur as each atom crosses the laser beam; the momentum lost in each decay is essentially distributed over all directions, so that there is a cumulative deflection that enables the selected isotope to be separately collected [2]. Nowadays, other methods of

doi:10.1088/978-0-7503-4191-2ch4

isotope separation, mostly non-optical and better suited to large-scale operations, have become more common.

4.2 Doppler cooling

In 1933, Frisch first demonstrated that radiation pressure could act on atoms by deflecting an atomic sodium beam with light from a lamp [3]. Much later, in 1975, the possibility of such an effect leading to the cooling of atoms by laser radiation was independently proposed by Hänsch and Schawlow [4], and also by Wineland and Dehmelt [5]; other developments then quickly followed [6–12]. The initial experiments on the topic involved the cooling of a cloud of Mg^{2+} ions confined to a Penning trap [13] and the cooling of trapped Ba^{+} ions [14].

Laser (or Doppler) cooling relates to the radiation force and its operation in atoms is described as follows. Electromagnetic radiation irradiates an atom with a frequency 'detuned to the red', i.e. the light frequency is slightly lower than a transition within the atom—the latter usually assumed to be a two-level system. Resonance conditions are met, resulting in photon absorption, when the atom is moving towards the light source—since the radiation becomes blue-shifted (relative to the atom) due to the Doppler effect. The momentum of the atom is then lowered by an amount equal to the momentum of the absorbed photon, $\hbar\mathbf{k}$. The excited atom soon relaxes back into the unexcited state via spontaneous emission. The recoil radiation that results is emitted in a random direction and, hence, it is highly unlikely to be equal to $\hbar\mathbf{k}$. Applying many absorption–emission events to an ensemble of atoms that is irradiated by two counter-propagating beams (a 1D set-up), the momentum from the recoil radiation averages to zero and the net momentum transferred to the atoms is $n_\phi\hbar\mathbf{k}$ in a direction opposite to atomic motion, producing a damping force. Therefore, the velocities of the atoms are reduced and the temperature is lowered. It is worth emphasising that the mechanism at work here is specifically a two-step process of photon absorption and emission; it is *not* scattering.

In the cycle of repeated absorption and spontaneous emission events, the mean reaction force resulting from the emissions is zero; hence the time-averaged recoil velocity $\langle\delta\mathbf{v}\rangle = 0$. Indeed, the mean velocity of the trapped atom itself becomes zero, i.e. $\langle\mathbf{v}\rangle = 0$. Nonetheless the quadratic mean $\langle|\mathbf{v}|^2\rangle$ need not vanish and, additionally, there are statistical fluctuations as the atom experiences the effects of individual photon absorption and emission events. Accordingly, since $\langle|\mathbf{v}|^2\rangle$ is non-zero, the atoms have a residual kinetic energy, as first observed by Bjorkholm *et al* [15]. This imposes a lower limit on the achievable extent of cooling, known as the *Doppler cooling limit*. An expression to determine the latter is written as $T_{\min} = \hbar\gamma/2k_B$ where $T_{\min}$ is the minimum temperature, k_B is the Boltzmann constant and γ is the natural linewidth of the atomic transition [16]. For example, an atomic gas at a pressure of 1 Torr, with a natural linewidth of 10 MHz, would give a limit of 50 μK. This expression was experimentally verified by Chang *et al* using metastable helium-4 gases in the Doppler regime [17]. By the adjunct use of a magneto-optical trap (MOT, see below), the limiting capabilities of atom cooling are, therefore, achieved at cryogenic temperatures.

Once the Doppler cooling limit is reached, diffusive motion of the ultracold atoms is observed [18–21]. There is a laser beam configuration—comprising, in three dimensions, three pairs of counter-propagating beams (figure 4.1)—that produces a loose assembly of diffusing ultracold atoms, sometimes known as *optical molasses*, so named because the beams may be thought to produce something that resembles a 'viscous fluid'. Initially used by Chu *et al* [18] to confine and cool sodium atoms, an important facet in the earliest experiments was the provision of a frequency chirp [22], i.e. a rapid shift in optical frequency with time. This frequency alteration counters the change in the Doppler shift as the atoms slow, enabling continuous resonance absorption of light by the atoms. In the late 80s, following the discovery that the Doppler cooling limit could be overcome [23, 24], it was recognised that other cooling mechanisms may occur in the system. This realisation led to the development of sub-Doppler laser cooling techniques such as Sisyphus cooling [25].

At a few degrees Kelvin, there is a trade-off between the limit of trapping time and the number of atoms that can be trapped; the atom number approximately scales with the square root of the sustainable cloud duration. The push towards lengthier trapping times continues to drive a progressive improvement in atom trapping performance, and single-atom lifetimes of up to ten minutes are now achievable [26].

A typical MOT device for trapping atoms is shown in figure 4.2. Its electromagnetic coils produce a quadrupolar magnetic field with a minimum at the intersection of the laser beams, such that the energy levels of an atom moving out of that region experience a Zeeman shift back towards resonance, resulting in the atom being pushed back inwards. (The alternative use of coupled electric and magnetic fields in a Penning trap only works for ions.) With three fibre-coupled laser inputs the MOT illustrated here can cool ^{85}Rb atoms to the recoil-temperature limit

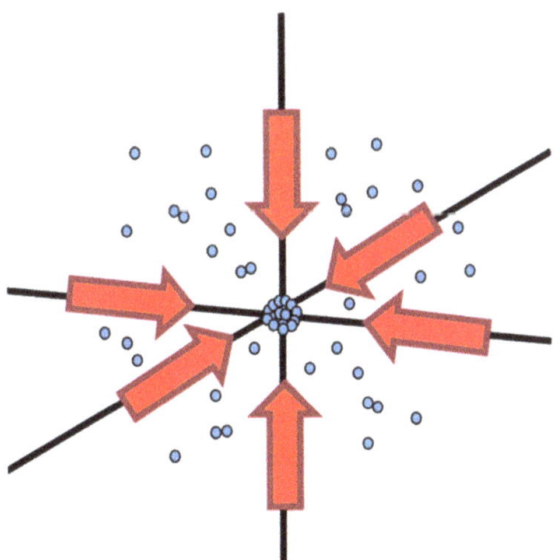

Figure 4.1. Schematic for the laser cooling of atoms within an optical molasses set-up. The red arrows denote pairs of counter-propagating beams and the circles signify atoms; a representation of a condensed cloud of atoms is centred at the intersection of the lasers.

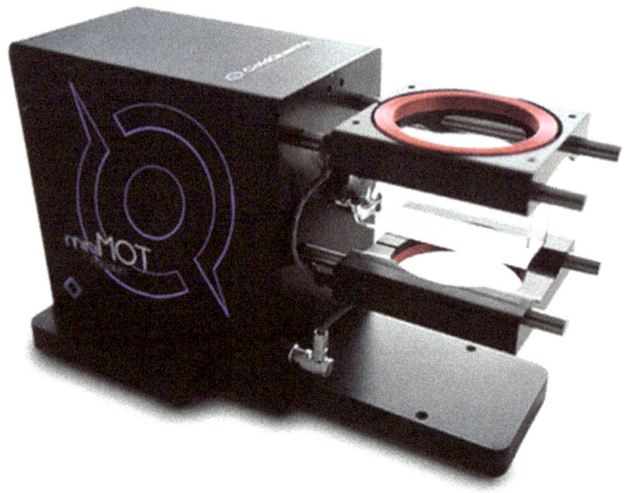

Figure 4.2. Compact magneto-optical trap for ultrahigh vacuum operation with Rb or Cs atoms. Image courtesy of ColdQuanta, by kind permission.

at around 10 μK, within microseconds, using input radiation suitably detuned from the D2 transition at 780.2 nm [27].

Doppler cooling requires the continuous absorption–emission cycle of a very large number of photons in a closed two-level transition. In practice, atomic systems rarely behave as completely two-level systems. For example, sodium atoms have two hyperfine ground sub-levels: for the purpose of driving transitions to achieve cooling, one is useful and the other is not. If spontaneous decay to unwanted states occurs, the two-level absorption–emission cycle will be compromised and ultracold temperatures may become unobtainable. In such a scenario, an effective two-level cycle is secured by re-pump lasers that excite the 'lost' population back into the cycle [28]. Unlike atoms, diatomic molecules contain vibrational and rotational states. As a result, application of the re-pump technique to these molecules is extremely difficult to implement, requiring an impractical number of re-pump lasers of various frequencies. Therefore, to achieve the laser cooling of diatomic molecules by any optical molasses set-up is scarcely viable. Although the production of ultracold molecules by laser cooling was first convincingly demonstrated over a decade ago using a configuration of lasers acting on strontium monofluoride [29–31]—and, more recently, yttrium oxide [32], strontium monohydroxide [33], ytterbium monofluoride [34] and calcium monofluoride [35, 36]—it has become apparent that no more than a dozen diatomic molecules have the internal structure amenable for similar laser cooling methods [37].

4.3 Bose–Einstein condensates

A combination of sub-Doppler laser cooling—the Nobel Prize-winning technique developed by Chu, Cohen-Tannoudji and Phillips—and evaporative cooling can produce

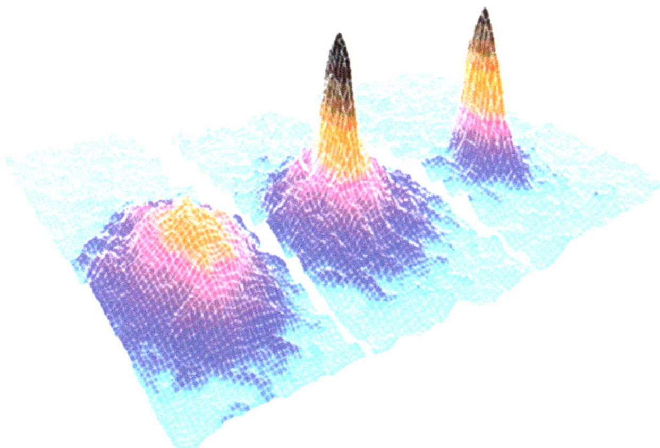

Figure 4.3. False-colour images displaying the velocity distribution for the cross-section of a cloud after the release of rubidium atoms from a trapping potential: (left) just above the phase transition, (centre) partly condensed, and (right) an almost pure condensate. This 'Bose–Einstein condensate' image has been obtained by the authors from the Wikimedia website, where it is stated to have been released into the public domain. It is included within this book on that basis.

a state of matter known as a Bose–Einstein condensate (BEC) [38–40], in which multiple ultracold atoms together enter a single, lowest energy quantum state. Such a state is only achievable for particles with a net spin that has an integer value (these are bosons). In 1995, the first BEC was produced when a vapour of rubidium-87 atoms condensed into a BEC at 170 nanokelvin [41], an achievement that led to the award of the Nobel Prize to Cornell and Wieman. The formation of a BEC requires that the ultracold sample has a de Broglie wavelength (h/p_a, where p_a is the mean atomic momentum associated with thermal motion and h is Planck's constant) that is larger than the average spacing between atoms, at a temperature below a critical limit. For an ideal (non-interacting) Bose gas, this critical temperature T_c is determined from the expression $T_c = 3.31\hbar^2 n_a^{2/3}/mk_B$, where m is the mass of each atom and n_a is the atom number density [42].

While a sub-Doppler temperature is achievable by the methods described above, evaporative cooling techniques [43–45] can further decrease the temperature so that the critical limit can be attained. Evaporative cooling works on the principle that, following confinement of the atoms in a magneto-optical trap, the lowering of the sides of the potential well (in which the trapped atoms reside) allows the most energetic atoms to escape. The result is that the remaining atoms contain lower kinetic energies and, thus, a lower temperature. Cooling by evaporation has the dual effect of increasing the density of the trapped atoms and lowering the temperature, both are crucial in the production of BECs. The formation of a BEC is illustrated in figure 4.3. Depending on the number of atoms involved, the lifetime of the condensate is typically on a millisecond timescale, but increasingly sophisticated experimentation now allows coherence to be sustained for several seconds.

References

[1] Janik G R, Cannon B D, Ogorzalek-Loo R and Bushaw B A 1989 Isotopically selective optical deflection of a krypton atomic beam *J. Opt. Soc. Am. B* **6** 1617–22

[2] Jackson D E and Davis J A 1985 Statistics of photodeflection *Transp. Theory Stat. Phys.* **14** 639–53

[3] Frisch O R 1933 Experimenteller nachweis des einstenschen strahlungsruckstosses *Z. Phys.* **86** 42–8

[4] Hänsch T W and Schawlow A L 1975 Cooling of gases by laser radiation *Opt. Commun.* **13** 68–9

[5] Wineland D J and Dehmelt H 1975 Proposed 10^{14} $\Delta\nu < \nu$ laser fluorescence spectroscopy on Tl^{+} mono-ion oscillator *Bull. Am. Phys. Soc.* **20** 637

[6] Letokhov V S, Minogin V G and Pavlik B D 1977 Cooling and capture of atoms and molecules by a resonant light field *Sov. Phys. JETP* **45** 698–705

[7] Wineland D J and Itano W M 1979 Laser cooling of atoms *Phys. Rev. A* **20** 1521–40

[8] Javanainen J 1980 Light-pressure cooling of trapped ions in three dimensions *Appl. Phys.* **23** 175–82

[9] Javanainen J and Stenholm S 1980 Broad band resonant light pressure II: cooling of gases *Appl. Phys.* **21** 163–7

[10] Letokhov V S and Minogin V G 1981 Laser radiation pressure on free atoms *Phys. Rep.* **73** 1–65

[11] Stenholm S 1986 The semiclassical theory of laser cooling *Rev. Mod. Phys.* **58** 699–739

[12] Phillips W D, Gould P L and Lett P D 1988 Cooling, stopping, and trapping atoms *Science* **239** 877–83

[13] Wineland D J, Drullinger R E and Walls F L 1978 Radiation-pressure cooling of bound resonant absorbers *Phys. Rev. Lett.* **40** 1639–42

[14] Neuhauser W, Hohenstatt M, Toschek P and Dehmelt H 1978 Optical-sideband cooling of visible atom cloud confined in parabolic well *Phys. Rev. Lett.* **41** 233–6

[15] Bjorkholm J E, Freeman R R, Ashkin A and Pearson D B 1980 Experimental observation of the influence of the quantum fluctuations of resonance-radiation pressure *Opt. Lett.* **5** 111–3

[16] Lett P D, Phillips W D, Rolston S L, Tanner C E, Watts R N and Westbrook C I 1989 Optical molasses *J. Opt. Soc. Am. B* **6** 2084–107

[17] Chang R, Hoendervanger A L, Bouton Q, Fang Y, Klafka T, Audo K, Aspect A, Westbrook C I and Clément D 2014 Three-dimensional laser cooling at the Doppler limit *Phys. Rev. A* **90** 063407

[18] Chu S, Hollberg L, Bjorkholm J E, Cable A and Ashkin A 1985 Three-dimensional viscous confinement and cooling of atoms by resonance radiation pressure *Phys. Rev. Lett.* **55** 48–51

[19] Phillips W D, Prodan J V and Metcalf H J 1985 Laser cooling and electromagnetic trapping of neutral atoms *J. Opt. Soc. Am. B* **2** 1751–67

[20] Sesko D, Fan C G and Wieman C E 1988 Production of a cold atomic vapor using diode-laser cooling *J. Opt. Soc. Am. B* **5** 1225–7

[21] Hodapp T W, Gerz C, Furtlehner C, Westbrook C I, Phillips W D and Dalibard J 1995 Three-dimensional spatial diffusion in optical molasses *Appl. Phys. B* **60** 135–43

[22] Ertmer W, Blatt R, Hall J L and Zhu M 1985 Laser manipulation of atomic beam velocities: demonstration of stopped atoms and velocity reversal *Phys. Rev. Lett.* **54** 996–9

[23] Lett P D, Watts R N, Westbrook C I, Phillips W D, Gould P L and Metcalf H J 1988 Observation of atoms laser cooled below the Doppler limit *Phys. Rev. Lett.* **61** 169–72

[24] Shevy Y, Weiss D S, Ungar P J and Chu S 1989 Bimodal speed distributions in laser-cooled atoms *Phys. Rev. Lett.* **62** 1118–21

[25] Dalibard J and Cohen-Tannoudji C 1989 Laser cooling below the Doppler limit by polarization gradients: simple theoretical models *J. Opt. Soc. Am. B* **6** 2023–45
[26] Schymik K-N, Pancaldi S, Nogrette F, Barredo D, Paris J, Browaeys A and Lahaye T 2021 Single atoms with 6000-second trapping lifetimes in optical-tweezer arrays at cryogenic temperatures *Phys. Rev. Appl.* **16** 034013
[27] Huang C, Kuan P-C and Lan S-Y 2018 Laser cooling of ^{85}Rb atoms to the recoil-temperature limit *Phys. Rev. A* **97** 023403
[28] Phillips W D 1998 Laser cooling and trapping of neutral atoms *Rev. Mod. Phys.* **70** 721–41
[29] Shuman E S, Barry J F, Glenn D R and DeMille D 2009 Radiative force from optical cycling on a diatomic molecule *Phys. Rev. Lett.* **103** 223001
[30] Shuman E S, Barry J F and DeMille D 2010 Laser cooling of a diatomic molecule *Nature* **467** 820–3
[31] Norrgard E B, McCarron D J, Steinecker M H, Tarbutt M R and DeMille D 2016 Submillikelvin dipolar molecules in a radio-frequency magneto-optical trap *Phys. Rev. Lett.* **116** 063004
[32] Hummon M T, Yeo M, Stuhl B K, Collopy A L, Xia Y and Ye J 2013 2D magneto-optical trapping of diatomic molecules *Phys. Rev. Lett.* **110** 143001
[33] Kozyryev I, Baum L, Matsuda K, Boerge H and Doyle J M 2016 Radiation pressure force from optical cycling on a polyatomic molecule *J. Phys. B: At. Mol. Opt. Phys.* **49** 134002
[34] Lim J, Almond J R, Trigatzis M A, Devlin J A, Fitch N J, Sauer B E, Tarbutt M R and Hinds E A 2018 Laser cooled YbF molecules for measuring the electron's electric dipole moment *Phys. Rev. Lett.* **120** 123201
[35] Zhelyazkova V, Cournol A, Wall T E, Matsushima A, Hudson J J, Hinds E A, Tarbutt M R and Sauer B E 2014 Laser cooling and slowing of CaF molecules *Phys. Rev. A* **89** 053416
[36] Anderegg L, Augenbraun B L, Bao Y, Burchesky S, Cheuk L W, Ketterle W and Doyle J M 2018 Laser cooling of optically trapped molecules *Nature Phys.* **14** 890–3
[37] McCarron D 2018 Laser cooling and trapping molecules *J. Phys. B: At. Mol. Opt. Phys.* **51** 212001
[38] Bose S N 1924 Plancks gesetz und lichtquantenhypothese *Z. Phys.* **26** 178–81
[39] Einstein A 1924 Quantentheorie des einatomigen idealen gases *Sitzungsber. Phys.-Math.* **Kl** 261–7
[40] Einstein A 1924 Quantentheorie des einatomigen idealen gases. Zweite abhandlung *Sitzungsber. Phys.-Math.* **Kl** 3–14
[41] Anderson M H, Ensher J R, Matthews M R, Wieman C E and Cornell E A 1995 Observation of Bose–Einstein condensation in a dilute atomic vapor *Science* **269** 198–201
[42] Bagnato V, Pritchard D E and Kleppner D 1987 Bose–Einstein condensation in an external potential *Phys. Rev. A* **35** 4354–8
[43] Hess H F 1986 Evaporative cooling of magnetically trapped and compressed spin-polarized hydrogen *Phys. Rev. B* **34** 3476–9
[44] Tommila T 1986 Cooling of spin-polarized hydrogen atoms trapped in magnetic-field minima *Europhys. Lett.* **2** 789–95
[45] Masuhara N, Doyle J M, Sandberg J C, Kleppner D, Greytak T J, Hess H F and Kochanski G P 1988 Evaporative cooling of spin-polarized atomic hydrogen *Phys. Rev. Lett.* **61** 935–8

IOP Publishing

Optical Nanomanipulation (Second Edition)

David L Andrews and David S Bradshaw

Chapter 5

Dielectric and metal nanoparticles: Rayleigh regime

Following Arthur Ashkin's development of the first optical trap for microparticles, based on radiation forces, several years later he created a technique now called optical tweezers, which relies on gradient forces to operate. It is the latter method that is the most useful for the trapping of nanoparticles, such a scheme is especially valuable for its ground-breaking ability to optically manipulate biological particles. While the trapping of microparticles has become routine for over 30 years, within the last two decades it has been demonstrated that various nanoscale semiconductor and dielectric particles can also be optically tweezered. In addition, plasmonic schemes have been developed to trap metal nanoparticles. The separation of chiral molecules by optical means is also achievable.

5.1 Arthur Ashkin and optical tweezers

From 1978 onwards, Ashkin demonstrated that neutral dielectric microparticles (and atoms) can be trapped with single-beam techniques that utilise radiation forces [1–3]. He also began to develop an ancillary means of optical trapping, based on the differential forces experienced by particles in different intensity regions of the laser beam; a mechanism that has become known as the gradient force [4]. In 1986, the first single-beam trap based on such a force was produced by Ashkin *et al* [5]—the set-up itself is now known as *optical tweezers*. Using an argon laser, it was shown possible for dielectric particles of those sizes ranging from 10 μm down to 25 nm to be stably trapped in water. The success of optically trapping particles with dimensions at the lower end of this scale represented the first achievement of optical nanomanipulation in the *Rayleigh-scattering regime*—a term signifying a particle diameter much smaller than the wavelength of the trapping light (i.e. $a \ll \lambda$), with scattering properties relating to the polarisability [6, 7].

doi:10.1088/978-0-7503-4191-2ch5

Figure 5.1. Arthur Ashkin on a video call in December 2018. He is the creator of the first optical tweezers technique, for which he was co-awarded the Nobel Prize in Physics in 2018. This 'Arthur Ashkin' image has been obtained by the authors from the Wikimedia website, where it was made available by Bengt Nyman under a CC BY-SA 2.0 licence. It is included within this book on that basis. It is attributed to Bengt Nyman.

Ashkin *et al* predicted that their new technique would 'open a new size regime to optical trapping encompassing macromolecules, colloids, small aerosols, and possibly biological particles.' His remark proved to be prophetic, since substances of all these types can indeed be trapped using present-day optical tweezer techniques. In fact, only a year after his ground-breaking studies, Ashkin's group was the first to use optical tweezers to trap and manipulate viruses, bacteria and single cells [8, 9]; building on this foundation, it is now even possible to trap red blood cells *in vivo* [10]. We discuss the optical manipulation of biomolecules in more details in chapter 7. It is interesting that the development of optical tweezers has been much more rapid than for the radiation pressure traps of previous years, stemming from the realisation by the scientific community that such a technique leads to a wide range of applications. By developing the optical tweezers, Ashkin (pictured in figure 5.1) once again proved to be the definitive pioneer of optical trapping and manipulation.

Optical tweezers work on the principle that a highly focused monochromatic continuous-wave (cw) laser beam [11, 12], typically with a Gaussian (TEM_{00} mode) intensity profile, can optically trap and manipulate micro- and nano-sized dielectric particles in a transparent medium. Although most optical tweezers are based on this cw type of laser, it is also now possible to use pulsed picosecond or femtosecond lasers with a fast pulse repetition rate, (typically in the GHz range), in order to exploit the much higher intensities, and hence forces, which they afford. In all such techniques, created by using a high-quality microscope objective, the particles are trapped within the narrow beam waist of the focused beam. While the gradient force attracts the particles laterally towards the high intensity part of the beam at its centre, the radiation force acts to 'push' the particles in the direction of beam propagation. However, the possibility of axial displacement from the focal region is countered by a diminution of intensity, due to the beam divergence, away from this region. The result is that the particle remains in a fixed axial position along the beam. Modern devices (for example, the BioPhotonics Workstation [13] of figure 5.2) operate under the control of computers, which are able to measure displacements and forces with high precision and accuracy.

An alternative to the single-beam configurations is the dual counter-propagating beam set-up, analogous to the first optical trap constructed in 1970 by Ashkin [14], in which the radiation forces are (to a large extent) balanced out by the two oppositely facing beams. This form of optical trapping was revolutionised by a fibre optics technique created in 1993 by Constable *et al* [15], who demonstrated that such a scheme may provide a number of enhanced features, including the ability to trap large cells. A few years later Zemanek *et al* designed a counter-propagating geometry, wherein an incident beam and its reflection interfere to produce a Gaussian standing wave trap, that operates at powers considerably smaller than optical tweezers when trapping nanoparticles [16]. However, while these forms of trapping require only two weakly focused beams, experimental complexities are added compared to the single beam trap, such as constraints in terms of the precision optical alignment.

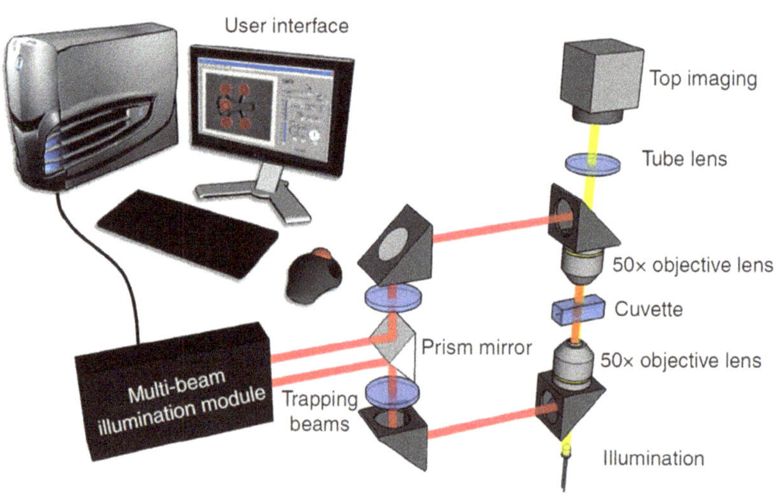

Figure 5.2. An example of a modern, computer-controlled optical trapping system. Known as the BioPhotonics Workstation, this set-up generates counter-propagating beam traps using a multi-beam illumination module. The top and bottom set of counter-propagating beams are imaged in the cuvette through opposing 50× objective lenses. The top imaging is fed real-time to the user-interface for intuitive optical manipulation. Reprinted by permission from Springer Nature: Nature. *Light: Science & Applications*, from reference [13], copyright 2016.

5.2 Optical trapping of nanoparticles

In the three decades that followed the initial devising of the optical tweezers methodology, experimental studies most commonly focused on the trapping of micro-sized particles such as beads and cells. However, the optical trapping of much smaller, nanoscale-dimension particles has developed into an even more rapidly growing field [17–20]. An effective deterrent to technical interest in trapping techniques for nanoparticles stems from the complexities in their implementation; the application of a sufficient optical force to a particle of nanoscale dimensions presents a challenge, especially when particles are so small that they experience only shallow intensity gradients even across a focused laser beam—see equations (3.7) and (3.8). The trapping of nanoparticles is much more readily achievable when at least one of the particle dimensions extends into the micron size range.

Dielectric and semiconductor nanoparticles. Examples here include bundles of carbon nanotubes [21, 22] (trapping of individual nanotubes remains a problematic task), semiconductor nanowires [23–25], nanorods [26, 27], 2D graphene flakes [28, 29] and nanosheets of boron nitride [30]; quantum dots have also been optically trapped [31, 32]. The process of angular orientation of an optically trapped nanorod, as the result of an optical torque defined by equation (3.9), is illustrated in figure 5.3.

In modelling terms, while a nanoparticle (such as a small molecule) can be treated within the electric dipole approximation—i.e. the particle is represented as a point electric dipole in an inhomogeneous electromagnetic field—most other nanostructures cannot be satisfactorily modelled in this way. Computational methods are used instead, such as: the *T*-matrix method [33–36], where non-spherical objects are modelled as clusters of small spheres; or the discrete dipole approximation [37],

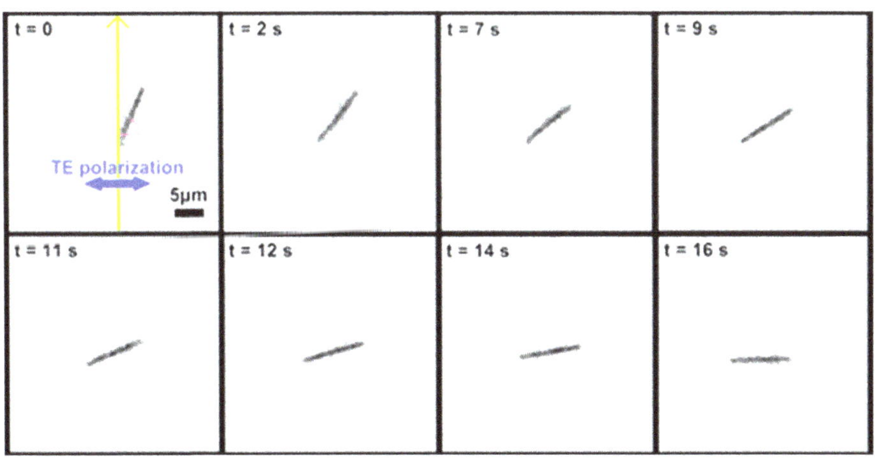

Figure 5.3. Sequential images showing the orientation of a nanorod (black line) into the direction of the electric field of the trapping beam (represented by the blue arrow in the $t = 0$ image); the green arrow denotes the propagation direction of the laser beam. The length of the nanorod is estimated as 10 μm and its diameter is 25 nm. Adapted with permission from reference [58]. Copyright 2012 American Chemical Society.

where the nanoparticle is treated as a collection of dipoles. The continued experimental and theoretical advancements in the optical trapping, manipulating and assembling of nanostructures should lead to the production of novel nano-electronic and nanophotonic devices in the near future. Furthermore, *laser refrigeration* has been demonstrated, in which an optically trapped nanocrystal immersed in water generates localised cooling of more than 10 °C below ambient conditions. This is achieved by harnessing the mechanism of anti-Stokes photoluminescence, in which laser excitation leads to blue-shifted emission that removes quanta of energy from thermally accessed states [38].

Metal nanoparticles. In the early days of optical trapping it was suggested that metallic objects could not be tweezed, since particles with high reflectivity are pushed out of the high-intensity region of a beam [2]. In 1994, however, Svoboda and Block experimentally showed that metallic Rayleigh particles can act in a manner analogous to their dielectric counterparts when optically trapped [39]—in fact, the larger polarisability of metal implies that the trapping forces will be greater. This finding spurred on progress in the optical manipulation of metal nanoparticles in liquids [18, 40, 41], and even optical control in air has recently been achieved [42, 43]. In contrast to standard non-resonant optical trapping, techniques have been developed based on the resonant excitation of free electrons in metal nanoparticles [44–46]; the oscillation in electron density that results is known as a *plasmon*. When the frequency of the trapping light matches the (usually gold or silver) nanoparticle's plasmon resonance, the optical response properties of the nanoscale metal object become significantly enhanced. Experimental exploration of the trapping of metal nanoparticles close to their resonance gained momentum following works by Pelton *et al* [47, 48] and Selhuber-Unkel *et al* [49], who optically trapped individual gold nanorods.

A related plasmonic scheme involves the trapping of nanoparticles within the gaps between plasmonic dipole antennas (a pair of metal nanorods spaced by a nanoscale gap); here it is the localised, evanescent near-field of the nanoantennas that enables the trapping of particles with dimensions as little as 10 nm [50–52]. Capitalising on this advance, nano-optical tweezers capable of three-dimensional manipulation of sub-100 nm dielectric objects are now possible. These traps are built by engineering a bowtie plasmonic aperture at the extremity of an optical fibre (figure 5.4) [53]. Excellent recent reviews on plasmonic optical tweezers can be found [54–57]. An explanation on the phenomena of surface plasmons is provided at the end of chapter 10.

5.3 Separation of chiral molecules

Chiral molecules have two non-superimposable mirror-image forms, in which each enantiomer can be identified through its rotation of linearly polarized light in opposite directions. Enantiomers typically behave as if they are physically identical, but they can interact differently with other biological materials or synthetic compounds that are chiral. Since the receptors in human biology often consist of chiral molecules, drug action mostly relates to a specified enantiomer. This has

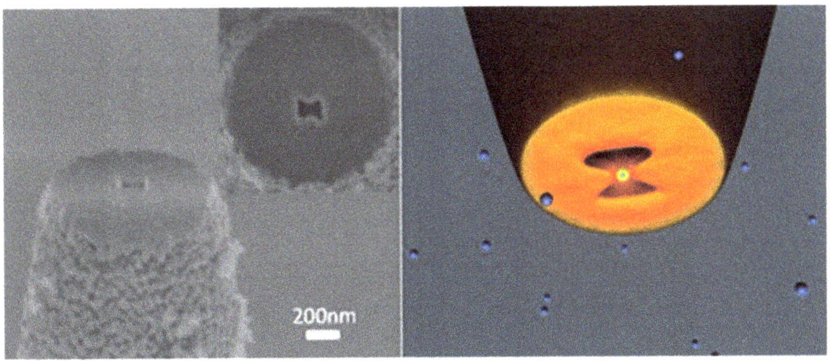

Figure 5.4. Electron beam microscopy image (left) shows the extremity of a plasmon nano-tweezer tip. The sketch on the right exhibits the trapping of a nanoparticle in a 'bowtie' plasmonic aperture. Reprinted by permission from Springer Nature: Nature. Nature Nanotechnology, from reference [53], copyright 2014.

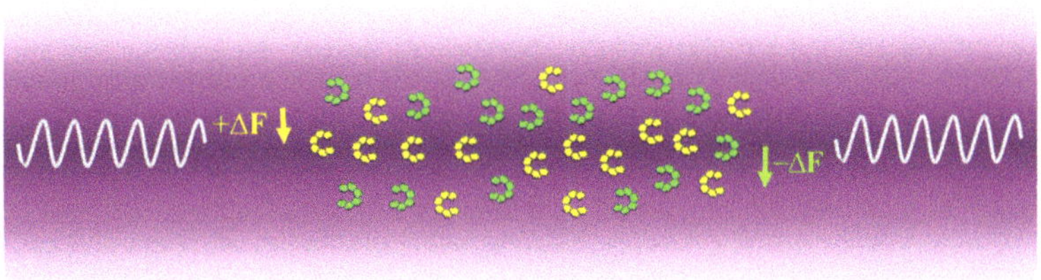

Figure 5.5. Diagram depicting the relative positions, within a circularly polarised trapping beam of Gaussian profile, of two typical enantiomers (S)-hexahelicene (yellow) and (R)-hexahelicene (green). Left-handed (S) molecules have a greater tendency, compared to the right-handed ones (R), to be positioned towards the centre of the Gaussian beam, as denoted by the chiral force **ΔF**; grey wavy line denotes the throughput beam.

prompted the pharmaceutical industry to develop a host of techniques to secure enantiopure products. Such methods, non-optical and time-consuming, can typically be cast in one of two distinct categories: synthetic mechanisms designed to produce a single enantiomer [59], or separation techniques to isolate distinct enantiomers from a racemic mixture [60].

In the last decade, non-contact optical methods have been devised to achieve this separation; these are mostly based on clever adaptions of optical traps [61–66]. While the gradient force of optical trapping is usually described within the electric dipole approximation, chirally sensitive interactions require that the magnetic transition dipole moments of the molecules are also considered. When the trapping laser is circularly polarised, it then emerges that the optical force acting on each enantiomer differs slightly to its oppositely handed counterpart—so that optical separation becomes possible. Such a difference in force is extremely small (with estimates in the 10^{-16} N range for a trapping beam intensity of 5×10^{11} W cm^{-2}) but experimentally viable. A potential physical system is illustrated in figure 5.5. An

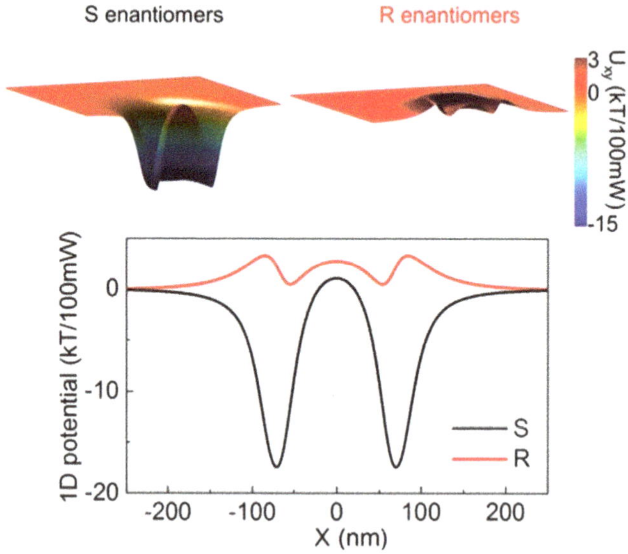

Figure 5.6. Two- and one-dimensional trapping potentials for enantiomers irradiated with circularly polarised light. The variation in the energy potential across the plasmonic aperture is shown, where the R- and S-enantiomers are positioned 20 nm above the aperture. The large difference in the potential well of an S-enantiomer (black curve) compared to an R-enantiomer (red curve) is the basis for the selective enantiomer trap. Reprinted with permission from reference [67]. Copyright 2016 American Chemical Society.

alternative separation scheme, based on an optical trap that uses a plasmonic aperture and circularly polarised light, has been proposed by Zhao *et al*: one enantiomer is trapped in a deep potential well and the other is repelled with a potential barrier (figure 5.6) [67]. A review article detailing the latest research on the optical manipulation of enantiomers, based on plasmonic schemes, is found in reference [68]. However, mainly due to the minuscule scale of the separation forces, major challenges need to be overcome if optical methods are ever to be viable on an industrial scale.

References

[1] Ashkin A 1978 Trapping of atoms by resonance radiation pressure *Phys. Rev. Lett.* **40** 729–32

[2] Ashkin A 1980 Applications of laser radiation pressure *Science* **210** 1081–8

[3] Ashkin A and Dziedzic J M 1985 Observation of radiation-pressure trapping of particles by alternating light beams *Phys. Rev. Lett.* **54** 1245–8

[4] Ashkin A and Gordon J P 1983 Stability of radiation-pressure particle traps: an optical Earnshaw theorem *Opt. Lett.* **8** 511–3

[5] Ashkin A, Dziedzic J M, Bjorkholm J E and Chu S 1986 Observation of a single-beam gradient force optical trap for dielectric particles *Opt. Lett.* **11** 288–90

[6] Harada Y and Asakura T 1996 Radiation forces on a dielectric sphere in the Rayleigh scattering regime *Opt. Commun.* **124** 529–41

[7] Malagnino N, Pesce G, Sasso A and Arimondo E 2002 Measurements of trapping efficiency and stiffness in optical tweezers *Opt. Commun.* **214** 15–24

[8] Ashkin A and Dziedzic J M 1987 Optical trapping and manipulation of viruses and bacteria *Science* **235** 1517–20

[9] Ashkin A, Dziedzic J M and Yamane T 1987 Optical trapping and manipulation of single cells using infrared laser beams *Nature* **330** 769–71

[10] Zhong M-C, Wei X-B, Zhou J-H, Wang Z-Q and Li Y-M 2013 Trapping red blood cells in living animals using optical tweezers *Nat. Commun.* **4** 1768

[11] Agate B, Brown C T A, Sibbett W and Dholakia K 2004 Femtosecond optical tweezers for in-situ control of two-photon fluorescence *Opt. Express* **12** 3011–7

[12] Gieseler J *et al* 2021 Optical tweezers—from calibration to applications: a tutorial *Adv. Opt. Photon.* **13** 74–241

[13] Villangca M J, Palima D, Banas A R and Gluckstad J 2016 Light-driven micro-tool equipped with a syringe function *Light: Sci. Appl.* **5** e16148

[14] Ashkin A 1970 Acceleration and trapping of particles by radiation pressure *Phys. Rev. Lett.* **24** 156–9

[15] Constable A, Kim J, Mervis J, Zarinetchi F and Prentiss M 1993 Demonstration of a fiber-optical light-force trap *Opt. Lett.* **18** 1867–9

[16] Zemánek P, Jonáš A, Šrámek L and Liška M 1999 Optical trapping of nanoparticles and microparticles by a Gaussian standing wave *Opt. Lett.* **24** 1448–50

[17] Maragò O M, Jones P H, Gucciardi P G, Volpe G and Ferrari A C 2013 Optical trapping and manipulation of nanostructures *Nat. Nanotechnol.* **8** 807–19

[18] Dienerowitz M, Mazilu M and Dholakia K 2008 Optical manipulation of nanoparticles: a review *J. Nanophoton.* **2** 021875

[19] Čižmár T D, Romero L C, Dholakia K and Andrews D L 2010 Multiple optical trapping and binding: new routes to self-assembly *J. Phys. B: At. Mol. Opt. Phys.* **43** 102001

[20] Spesyvtseva S E S and Dholakia K 2016 Trapping in a material world *ACS Photonics* **3** 719–36

[21] Maragò O M *et al* 2008 Optical trapping of carbon nanotubes *Phys. E* **40** 2347–51

[22] Mishra A, Clayton K, Velasco V, Williams S J and Wereley S T 2016 Dynamic optoelectric trapping and deposition of multiwalled carbon nanotubes *Microsys. Nanoeng.* **2** 16005

[23] Pauzauskie P J, Radenovic A, Trepagnier E, Shroff H, Yang P and Liphardt J 2006 Optical trapping and integration of semiconductor nanowire assemblies in water *Nat. Mater.* **5** 97–101

[24] van der Horst A, Campbell A I, van Vugt L K, Vanmaekelbergh D A M, Dogterom M and Blaaderen A V 2007 Manipulating metal-oxide nanowires using counter-propagating optical line tweezers *Opt. Express* **15** 11629–39

[25] Roder P B, Manandhar S, Smith B E, Zhou X, Shutthanandan V S and Pauzauskie P J 2015 Photothermal superheating of water with ion-implanted silicon nanowires *Adv. Opt. Mater.* **3** 1362–7

[26] Yu T, Cheong F-C and Sow C-H 2004 The manipulation and assembly of CuO nanorods with line optical tweezers *Nanotechnology* **15** 1732–6

[27] Kumar A, Kim J, Lahlil K, Julie G, Nic Chormaic S, Kim J, Gacoin T and Fick J 2020 Optical trapping and orientation-resolved spectroscopy of europium-doped nanorods *J. Phys. Photonics* **2** 025007

[28] Maragò O M *et al* 2010 Brownian motion of graphene *ACS Nano* **4** 7515–23

[29] Twombly C W, Evans J S and Smalyukh I I 2013 Optical manipulation of self-aligned graphene flakes in liquid crystals *Opt. Express* **21** 1324–34

[30] Donato M G, Messina E, Foti A, Smart T J, Jones P H, Iatì M A, Saija R, Gucciardi P G and Maragò O M 2018 Optical trapping and optical force positioning of two-dimensional materials *Nanoscale* **10** 1245–55

[31] Jauffred L, Richardson A C and Oddershede L B 2008 Three-dimensional optical control of individual quantum dots *Nano Lett.* **8** 3376–80

[32] Bendix P M, Jauffred L, Norregaard K and Oddershede L B 2014 Optical trapping of nanoparticles and quantum dots *IEEE J. Sel. Top. Quant. Electron.* **20** 15–26

[33] Borghese F, Denti P, Saija R and Iatì M A 2007 Optical trapping of nonspherical particles in the T-matrix formalism *Opt. Express* **15** 11984–98

[34] Nieminen T A, Loke V L Y, Stilgoe A B, Heckenberg N R and Rubinsztein-Dunlop H 2011 T-matrix method for modelling optical tweezers *J. Mod. Opt.* **58** 528–44

[35] Nieminen T A, du Preez-Wilkinson N, Stilgoe A B, Loke V L Y, Bui A A M and Rubinsztein-Dunlop H 2014 Optical tweezers: theory and modelling *J. Quant. Spectrosc. Radiat. Transfer* **146** 59–80

[36] Patti F, Saija R, Denti P, Pellegrini G, Biagioni P, Iatì M A and Maragò O M 2019 Chiral optical tweezers for optically active particles in the T-matrix formalism *Sci Rep.* **9** 29

[37] Draine B T and Flatau P J 1994 Discrete-dipole approximation for scattering calculations *J. Opt. Soc. Am. A* **11** 1491–9

[38] Roder P B, Smith B E, Zhou X, Crane M J and Pauzauskie P J 2015 Laser refrigeration of hydrothermal nanocrystals in physiological media *Proc. Natl Acad. Sci. USA* **112** 15024–9

[39] Svoboda K and Block S M 1994 Optical trapping of metallic Rayleigh particles *Opt. Lett.* **19** 930–2

[40] Bosanac L, Aabo T, Bendix P M and Oddershede L B 2008 Efficient optical trapping and visualization of silver nanoparticles *Nano Lett.* **8** 1486–91

[41] Hajizadeh F and Reihani S N 2010 Optimized optical trapping of gold nanoparticles *Opt. Express* **18** 551–9

[42] Jauffred L, Taheri S M-R, Schmitt R, Linke H and Oddershede L B 2015 Optical trapping of gold nanoparticles in air *Nano Lett.* **15** 4713–9

[43] Arita Y, Tkachenko G, McReynolds N, Marro N, Edwards W, Kay E R and Dholakia K 2018 Optical trapping of ultrasmooth gold nanoparticles in liquid and air *APL Photonics* **3** 070801

[44] Urban A S, Carretero-Palacios S, Lutich A A, Lohmuller T, Feldmann J and Jackel F 2014 Optical trapping and manipulation of plasmonic nanoparticles: fundamentals, applications, and perspectives *Nanoscale* **6** 4458–74

[45] Lehmuskero A, Johansson P, Rubinsztein-Dunlop H, Tong L and Käll M 2015 Laser trapping of colloidal metal nanoparticles *ACS Nano* **9** 3453–69

[46] Ivinskaya A, Petrov M I, Bogdanov A A, Shishkin I, Ginzburg P and Shalin A S 2017 Plasmon-assisted optical trapping and anti-trapping *Light: Sci. Appl.* **6** e16258

[47] Pelton M, Liu M, Kim H Y, Smith G, Guyot-Sionnest P and Scherer N F 2006 Optical trapping and alignment of single gold nanorods by using plasmon resonances *Opt. Lett.* **31** 2075–7

[48] Toussaint K C, Liu M, Pelton M, Pesic J, Guffey M J, Guyot-Sionnest P and Scherer N F 2007 Plasmon resonance-based optical trapping of single and multiple Au nanoparticles *Opt. Express* **15** 12017–29

[49] Selhuber-Unkel C, Zins I, Schubert O, Sönnichsen C and Oddershede L B 2008 Quantitative optical trapping of single gold nanorods *Nano Lett.* **8** 2998–3003
[50] Grigorenko A N, Roberts N W, Dickinson M R and Zhang Y 2008 Nanometric optical tweezers based on nanostructured substrates *Nat. Photonics* **2** 365–70
[51] Righini M, Ghenuche P, Cherukulappurath S, Myroshnychenko V, García de Abajo F J and Quidant R 2009 Nano-optical trapping of Rayleigh particles and Escherichia coli bacteria with resonant optical antennas *Nano Lett.* **9** 3387–91
[52] Zhang W, Huang L, Santschi C and Martin O J F 2010 Trapping and sensing 10 nm metal nanoparticles using plasmonic dipole antennas *Nano Lett.* **10** 1006–11
[53] Berthelot J, Acimovic S S, Juan M L, Kreuzer M P, Renger J and Quidant R 2014 Three-dimensional manipulation with scanning near-field optical nanotweezers *Nat. Nanotechnol.* **9** 295–9
[54] Huang J-S and Yang Y-T 2015 Origin and future of plasmonic optical tweezers *Nanomaterials* **5** 1048–65
[55] Bradac C 2018 Nanoscale optical trapping: a review *Adv. Opt. Mat.* **6** 1800005
[56] Kotsifaki D G and Chormaic S N 2019 Plasmonic optical tweezers based on nanostructures: fundamentals, advances and prospects *Nanophoton.* **8** 1227–45
[57] Zhang Y, Min C, Dou X, Wang X, Urbach H P, Somekh M G and Yuan X 2021 Plasmonic tweezers: for nanoscale optical trapping and beyond *Light: Sci. Appl.* **10** 59
[58] Kang P, Serey X, Chen Y-F and Erickson D 2012 Angular orientation of nanorods using nanophotonic tweezers *Nano Lett.* **12** 6400–7
[59] Rouf A and Taneja S C 2014 Synthesis of single-enantiomer bioactive molecules: a brief overview *Chirality* **26** 63–78
[60] Scriba G K E 2012 Chiral recognition mechanisms in analytical separation sciences *Chromatographia* **75** 815–38
[61] Cameron R P, Barnett S M and Yao A M 2012 Optical helicity, optical spin and related quantities in electromagnetic theory *New J. Phys.* **14** 053050
[62] Canaguier-Durand A, Hutchison J A, Genet C and Ebbesen T W 2013 Mechanical separation of chiral dipoles by chiral light *New J. Phys.* **15** 123037
[63] Bradshaw D S and Andrews D L 2014 Chiral discrimination in optical trapping and manipulation *New J. Phys.* **16** 103021
[64] Bradshaw D S and Andrews D L 2015 Laser optical separation of chiral molecules *Opt. Lett.* **40** 677–80
[65] Bradshaw D S and Andrews D L 2015 Electromagnetic trapping of chiral molecules: orientational effects of the irradiating beam *J. Opt. Soc. Am. B* **32** B25–31
[66] Ying X, Rui G, Zou S, Gu B, Zhan Q and Cui Y 2021 Synthesis of multiple longitudinal polarization vortex structures and its application in sorting chiral nanoparticles *Opt. Express* **29** 19001–14
[67] Zhao Y, Saleh A A E and Dionne J A 2016 Enantioselective optical trapping of chiral nanoparticles with plasmonic tweezers *ACS Photonics* **3** 304–9
[68] Kakkanattu A, Eerqing N, Ghamari S and Vollmer F 2021 Review of optical sensing and manipulation of chiral molecules and nanostructures with the focus on plasmonic enhancements *Opt. Express* **29** 12543–79

IOP Publishing

Optical Nanomanipulation (Second Edition)

David L Andrews and David S Bradshaw

Chapter 6

Larger nanoparticles: Lorenz–Mie regime and beyond

The optomechanical response of sub-micron particles with dimensions similar to the optical wavelength differs significantly from the effects exhibited by smaller nanoparticles. Here, Mie scattering theory is used to describe the response of particles with spherical symmetry; for less symmetric particles a treatment using a more intricate Lorenz–Mie theory is required, with a ray optics approach coming into play for transparent particles. Among a host of interesting research on such systems, optical levitation represents just one, particularly striking example. Thermal complications may be a challenge when particles are suspended in liquid media but, in other respects, they could be harnessed in a technique known as photophoresis. The generation of structured optical traps may enable particle sorting in fractionation methods.

6.1 Mie scattering

For particles with a major diameter a on the scale of the wavelength ($a \sim \lambda$), Mie scattering theory is commonly applied [1, 2]. This is a classical electrodynamical representation that can model the interaction between arbitrarily shaped beams and particles of spherical symmetry. A generalised Lorenz–Mie formalism represents an extension to particles of lower symmetry, or asymmetric species [3, 4]. In fact, transparent particles in the micron range dominate much of the field of optical trapping; such microparticles ($a \gg \lambda$) are usually modelled in a ray optics approach. In particles of this size, the trapping light undergoes internal refraction, leading to an optical force that acts towards the beam centre [5]. This mechanism is another example of the radiation force, since the optical field can be described in terms of rays that transfer linear momentum to the particle following Mie (inelastic) scattering. The mechanism for this description, shown on figure 6.1, is described as follows.

doi:10.1088/978-0-7503-4191-2ch6

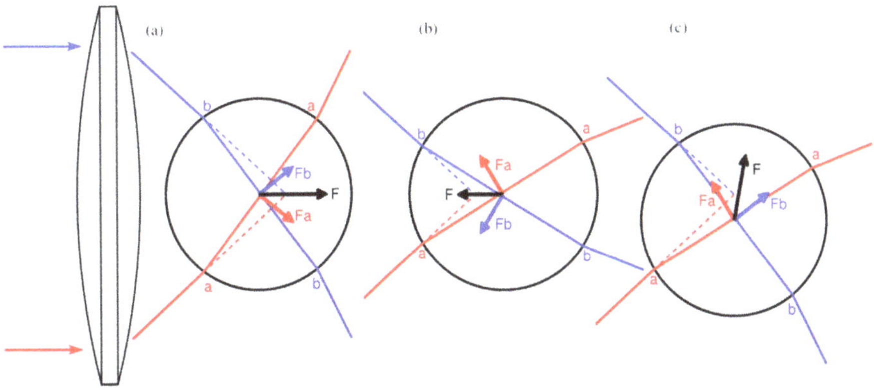

Figure 6.1. Qualitative view of the trapping of a dielectric sphere, showing that the refraction of a typical pair of rays a and b of the trapping beam result in forces, F_a and F_b respectively, whose sum is always restoring towards equilibrium for axial and transverse displacements of the sphere from the trap focus, located at the intersection of the dashed lines. Cases (a) and (b) show the axial restoring force acting on a particle that is axially displaced from the focal plane. Similarly, (c) shows the lateral restoring force upon the particle towards the beam focus. Reproduced from reference [6].

The total beam can be decomposed into individual rays—each with an appropriate intensity, direction and polarisation—that propagates in straight lines within uniform media of refractive index n_ω. Each ray may change direction when it reflects or refracts, and it changes polarisation at dielectric interfaces according to the Fresnel equations (which describe the behaviour of light when it traverses the interface between media of different refractive indices). The refraction of light rays at the particle surfaces results in a transfer of momentum from the trapping laser to the particle; the rate of delivery determines the overall trapping force [6, 7]. In quantum terms, these larger particles experience a propelling force towards the beam centre through the cumulative result of the directional differences in momentum between each photon entering and exiting the particle. Once again, it is the overall rate of such scattering events that determines the magnitude of the optical force.

6.2 Thermal effects in optical trapping

In thermal equilibrium, the natural Brownian motion of an optically trapped particle is confined to a finite trapping region. The probability density, N', for a particle position within the trap can be established from Boltzmann statistics, so that in one dimension;

$$N' \propto \mathrm{e}^{-\frac{\Delta E(x)}{k_B T}}, \tag{6.1}$$

where T represents the absolute temperature. It is interesting that N' tends to a position-independent constant for high temperatures or low potential energies, which physically represents a loss of a viable optical trap. Following a Taylor series expansion of equation (6.1), the leading term for a local position within the trap is given as;

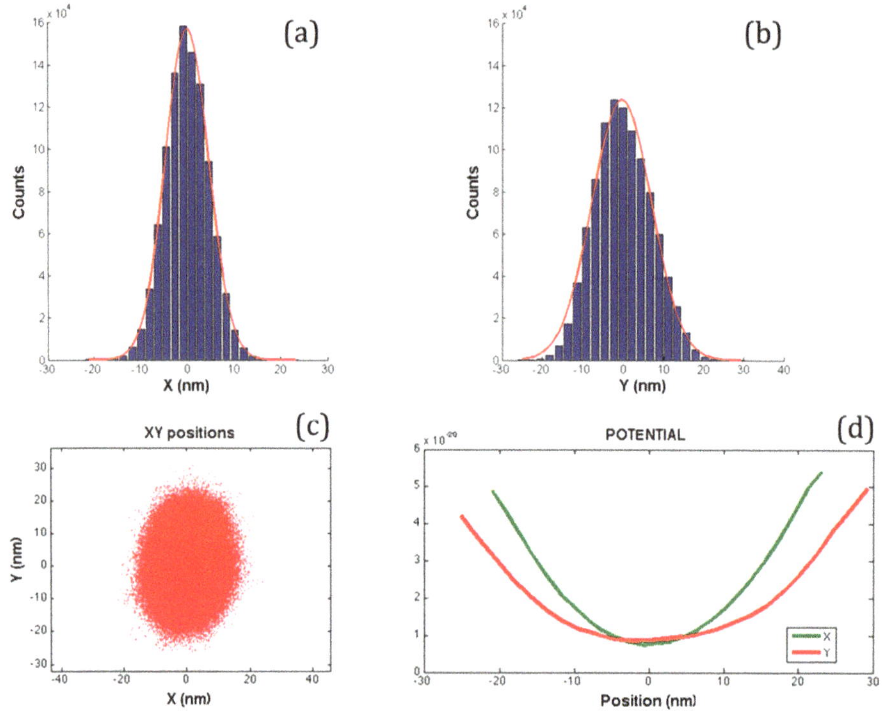

Figure 6.2. Histograms of (a) x-axis position and (b) y-axis position of a particle within an optical trap, and (c) a scatterplot of the particle position within the x,y-plane; all are determined by a repeated series of position measurements. (d) Potential energy curve (in Joules) for the optical trap in the x- and y-directions. The ellipsoid shape in (c) is due to disparity between the potential wells and corresponding particle distributions in the x- and y-directions. Images courtesy of Mario Montes Usategui and Estela Martín Badosa, Universitat de Barcelona.

$$\Delta N' \propto -\frac{\Delta E(x)}{k_{\mathrm{B}}T}, \tag{6.2}$$

which is analogous to Curie's law. It is usual for the trapping potential to be modelled as a harmonic potential, meaning that $\Delta E(x) \sim \frac{1}{2}k_x x^2$ where k_x is a force constant that relates to the stiffness of the trap. The latter can be expressed as $k_x \sim k_{\mathrm{B}}T/\sigma^2{}_x$ with σ_x as the variance of a Gaussian distribution curve. Inserting the last two formulae into equation (6.2), the following is determined;

$$\Delta N' \propto -\frac{x^2}{2\sigma_x^2} \tag{6.3}$$

Here, the drag due to the viscosity of the medium is not considered; this would require additional terms that are detailed elsewhere [8]. On taking a statistically significant number of position measurements, the histogram of particle positions (typically of the form shown in figure 6.2) has a similar shape to the spatial probability density of equation (6.1). As expected, the position of the particles is most probable at the centre of the beam where the potential energy is lowest. These three expressions equally apply to optical trapped nanoparticles. However, particles of such size have smaller polarisabilities (and, hence, smaller ΔE), which results in less variation in the probability density, and a shallower optical trap.

The term 'photophoresis' signifies another, more directed form of optically induced motion. Here, microscopic motion occurs as a result of local heating produced by light absorption [9]. It is, in fact, the same effect that is responsible for the rotating vanes in a 'Crookes radiometer', illustrated in figure 6.3—although the effect, whose discovery is correctly ascribed to Crookes, does not operate by the mechanism originally supposed [10]. As proven by Reynolds five years later [11], motion occurs as a result of gas molecules experiencing inelastic collisions on the heated surface, removing thermal energy at the expense of imparting kinetic energy and momentum to the heated material.

Figure 6.3. In an air-filled glass bulb, a vane rotates around a slender spindle as a result of the black side of each vane absorbing more light than its reverse side, which is silvered. Molecules of air colliding with the dark side acquire some of this thermal energy and bounce off at greater speed. This imparts a reverse momentum to the surface, and hence an angular momentum that sustains rotary motion. Photograph by David L Andrews.

In the nanoscale domain, such a radiometric effect can be produced by laser light. Although this may present a significant complication in radiation pressure applications, (to the point where the radiation force can be completely impaired), it can also be utilised in a tailored way that, for example, takes advantage of the wavelength dependence of absorption. Spherical nanoparticles can experience a photophoretic force that exceeds the radiation pressure force by a factor of up to $c/3v$, where v is the magnitude of the mean velocity of the gas molecules, depending on the nature of the gas and its temperature [12]. The upper bound on this ratio, which can be approached if there is negligible thermal conductance within the particle, is given by $(m/27k_{\mathrm{B}}T)^{1/2}c$, where m is the molecular mass of the surrounding fluid and T is the absolute temperature. Typically, the photophoretic effect in a gas can exceed that of radiation pressure by a factor of 10^6 or more; in a transparent liquid the ratio may be an order of magnitude or two less, though still easily sufficient to outweigh radiation forces. However, it is worth emphasizing that this effect is only operative when the nanoparticles strongly absorb the laser radiation, meaning that the laser wavelength has to lie well within an absorption band.

Photophoresis can, for example, be used to guide carbon nanocluster assembles suspended in air [12, 13]. Although much less understood [14], the same principle can be used for inducing guided motion to particles in liquids; although higher powers are required for the necessary convection currents, and Stokes drag has also to be considered [15].

6.3 Optical levitation

One of the earliest techniques developed by Ashkin in the formative years of laser trapping is optical levitation, whereby a particle is suspended in air (or a high vacuum) [16]. Here, the force of gravity, i.e. the particle mass m multiplied by the acceleration due to gravity g, is opposed by the radiation force of a vertically directed focused beam. The Maxwell–Bartoli formula of equation (3.1), which is generally geared towards large scale illumination of an inclined plane, can be readily modified to account for the realities of optical levitation. Assuming a particle of almost spherical shape, and dimensions smaller than the focused beam width, we interpret the light irradiance acting on the under surface of the particle as being only a fraction of the beam intensity. Moreover, a reflective particle will reflect light at various angles according to the positions at which each photon hits the surface, so most of the light is not reflected back towards its source. Assuming that the majority of the laser intensity at the focus lies within the beam-width parameter w_0, and for simplicity neglecting the variation of intensity across that region, by use of simple geometry the following force equation is found;

$$mg = \left(\frac{W}{c}\right)\left(\frac{a}{2w_0}\right)^2\left(1 + \frac{R}{3}\right), \tag{6.4}$$

where R is the reflectivity of the levitated particle and W is the power of the beam—averaged over time, if a rapidly pulsed source is deployed. This expression can be re-arranged to produce the following;

$$W = \frac{4\pi r w_0^2(\rho - \rho_0)cg}{R + 3} \tag{6.5}$$

Here, $\rho - \rho_0$ is the difference between the density of the particle and a supporting medium (as commonly applied to particles in liquid suspension). For example, using the values $2w_0 = 25$ μm, $\rho - \rho_0 = 50$ kg m^{-3} and $R = 0.6$, it emerges that a reflective particle with a diameter of 20 μm can be levitated by a laser with a power greater than 0.8 mW. Once aloft, the particle can be freely manipulated by simply moving the beam. However, unlike optical tweezers, this technique cannot be considered all-optical since it requires gravitational forces to function. Levitation of this type was initially demonstrated with a small transparent glass sphere [17] and, soon after, it was also achieved with oil and water droplets [18]. In more recent times, the optical levitation of a microdroplet containing a single quantum dot [19], and the suspension of a particle within a Bessel (non-diffracting) light beam [20], have been demonstrated. In aqueous suspension, it has also been shown that when 1–5 μm gold particles are levitated at the focus of a linearly polarised Gaussian beam, they can migrate to off-axis positions and spin as a result of a force differential across them [21]. Furthermore, it has been proposed that optically levitating a small curved mirror will enable its mechanically isolated, non-contact support to be achieved, and thereby utilised for ultrasensitive motion detection [22].

Current research on optical levitation often operates on nanoparticles; an example is illustrated by figure 6.4, in which a nanodiamond is optically levitated [23]. Such systems can act as a novel experimental platform for the exploration of microscopic thermodynamics and non-equilibrium physics [24, 25]. This led to the first observation of ballistic Brownian motion [26]. While optical manipulation techniques, such as these, usually involves particles within a suspending medium, there has been renewed interest in configurations in a high vacuum. This development is motivated by potential entry into the quantum regime, due to suppressed interactions with the surroundings. Such a goal, now considered to be within reach [25], has propelled the advancement of the field of levitated optomechanics [27, 28]. Now a vast subject area, some of the many highlights include the ability for ultrasensitive force sensing in the 10^{-21} N range [29] and for the controlled rotation of levitated silicon nanorods with elliptically polarised light [30]. Related to the latter, but with use of a circularly polarised beam, a 100 nm diameter silica nanosphere [31] and a 170 nm diameter silica nanodumbbell [32] have been spun at speeds beyond 1 billion rotations per second. These are the highest ever recorded rotational frequencies of a mechanical object; by way of comparison, the fans of an engine in a fighter jet typically spin at less than 1000 rotations per second. (It should be noted here that this technique differs to the far less efficient systems involving linearly polarised light, in which a nanorod simply aligns with an experimentally controlled, constantly rotating light polarisation [33].)

6.4 Optical fractionation

Known as *optical fractionation*, one way to achieve microfluidic sorting is with an optical trapping array (figure 6.5). The latter is discussed, in more detail, in chapter 8;

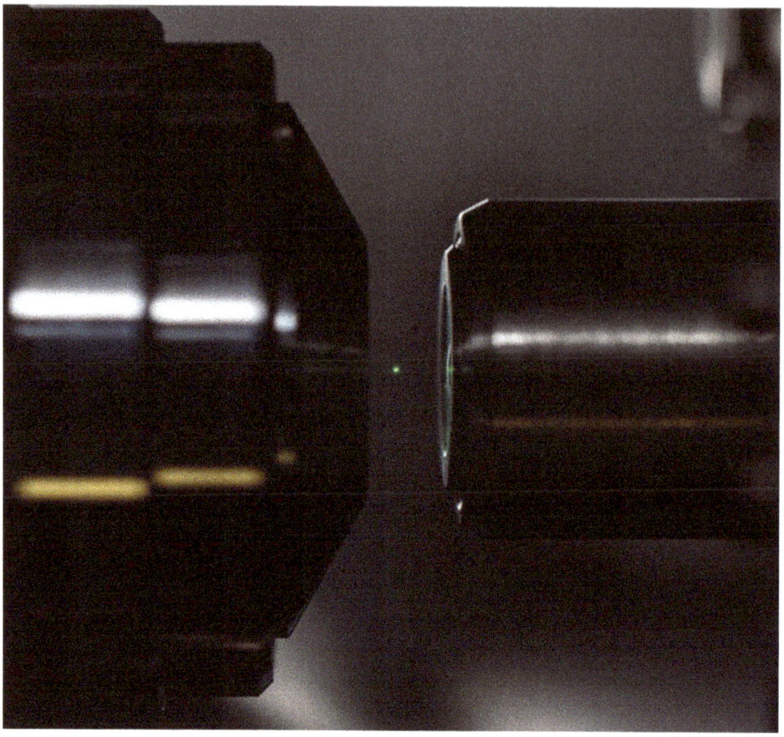

Figure 6.4. Photograph of a levitated nanodiamond illuminated by a green excitation beam. Reprinted by permission from Springer Nature: Nature. *Nature Photonics*, from reference [23], copyright 2015.

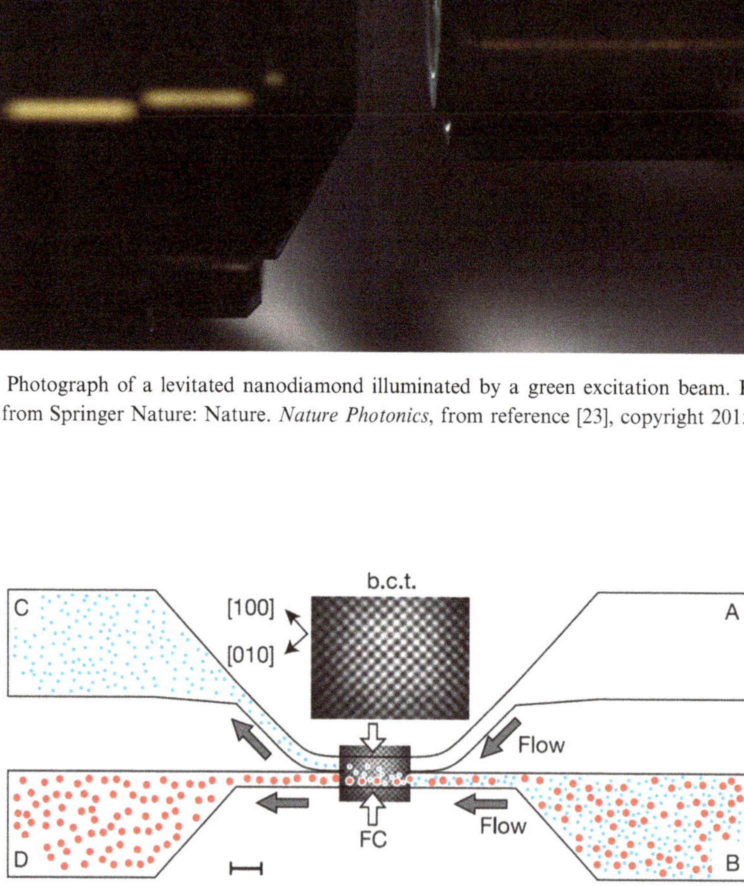

Figure 6.5. The concept of optical fractionation. Without an actuator all particles from chamber B will flow into chamber D, as a result of the non-turbulence of microfluidic flow. On introduction of an optical lattice—in this case, a body-centred tetragonal (b.c.t.) lattice—into the fractionation chamber (FC), one type of particle is selectively pushed into the upper flow field; chamber A relates to a 'blank' flow stream and the scale bar represents 40 μm. The reconfigurability of the optical lattice allows for dynamic updating of selection criteria. Reprinted by permission from Springer Nature: Nature. *Nature*, from reference [34], copyright 2003.

in short, it involves an intensity pattern that may have specifically designed spatial variations. Optical forces in such patterns depend on the particle shape, size and refractive index [34]. Many other techniques have been proposed, both with and without the use of a fluidic flow [35–41]. Some of these methods find application in the sorting of biological materials; they represent part of a much broader field of optical nanomanipulation, which is the subject of the next chapter.

References

[1] Mie G 1908 Beiträge zur Optik trüber Medien, speziell kolloidaler Metallösungen *Ann. Phys.* (Berlin) **330** 377–445

[2] Zangwill A 2012 *Modern Electrodynamics* (Cambridge: Cambridge University Press) p. 787

[3] Gouesbet G 2009 Generalized Lorenz–Mie theories, the third decade: a perspective *J. Quant. Spectrosc. Radiat. Transfer* **110** 1223–38

[4] Gouesbet G 2014 Latest achievements in generalized Lorenz–Mie theories: a commented reference database *Ann. Phys.* (Berlin) **526** 461–89

[5] Dholakia K, Reece P and Gu M 2008 Optical micromanipulation *Chem. Soc. Rev.* **37** 42–55

[6] Čižmár T D, Romero L C, Dholakia K and Andrews D L 2010 Multiple optical trapping and binding: new routes to self-assembly *J. Phys. B: At. Mol. Opt. Phys.* **43** 102001

[7] Ashkin A 1992 Forces of a single-beam gradient laser trap on a dielectric sphere in the ray optics regime *Biophys. J.* **61** 569–82

[8] Mas J, Farre A, Cuadros J, Juvells I and Carnicer A 2011 Understanding optical trapping phenomena: a simulation for undergraduates *IEEE Trans. Educ.* **54** 133–40

[9] Jonáš A and Zemánek P 2008 Light at work: the use of optical forces for particle manipulation, sorting, and analysis *Electrophoresis* **29** 4813–51

[10] Crookes W 1874 XV. On attraction and repulsion resulting from radiation *Philos. Trans. R. Soc.* **164** 501–27

[11] Reynolds O 1879 XVIII. On certain dimensional properties of matter in the gaseous state *Philos. Trans. R. Soc.* **170** 727–845

[12] Shvedov V G, Desyatnikov A S, Rode A V, Krolikowski W and Kivshar Y S 2009 Optical guiding of absorbing nanoclusters in air *Opt. Express* **17** 5743–57

[13] Desyatnikov A S, Shvedov V G, Rode A V, Krolikowski W and Kivshar Y S 2009 Photophoretic manipulation of absorbing aerosol particles with vortex beams: theory versus experiment *Opt. Express* **17** 8201–11

[14] Soong C Y, Li W K, Liu C H and Tzeng P Y 2010 Theoretical analysis for photophoresis of a microscale hydrophobic particle in liquids *Opt. Express* **18** 2168–82

[15] Flores-Flores E, Torres-Hurtado S, Páez R, Ruiz U, Beltrán-Pérez G, Neale S, Ramirez-San-Juan J and Ramos-García R 2015 Trapping and manipulation of microparticles using laser-induced convection currents and photophoresis *Biomed. Opt. Express* **6** 4079–87

[16] Ashkin A and Dziedzic J M 1976 Optical levitation in high vacuum *Appl. Phys. Lett.* **28** 333–5

[17] Ashkin A and Dziedzic J M 1971 Optical levitation by radiation pressure *Appl. Phys. Lett.* **19** 283–5

[18] Ashkin A and Dziedzic J M 1975 Optical levitation of liquid drops by radiation pressure *Science* **187** 1073–5

[19] Minowa Y, Kawai R and Ashida M 2015 Optical levitation of a microdroplet containing a single quantum dot *Opt. Lett.* **40** 906–9

[20] Garcés-Chávez V, Roskey D, Summers M D, Melville H, McGloin D, Wright E M and Dholakia K 2004 Optical levitation in a Bessel light beam *Appl. Phys. Lett.* **85** 4001–3

[21] Liang Y, Yan S, Wang Z, Yao B and Lei M 2021 Off-axis optical levitation and transverse spinning of metallic microparticles *Photon. Res.* **9** 2144–51

[22] Guccione G, Hosseini M, Adlong S, Johnsson M T, Hope J, Buchler B C and Lam P K 2013 Scattering-free optical levitation of a cavity mirror *Phys. Rev. Lett.* **111** 183001

[23] Neukirch L P, von Haartman E, Rosenholm J M and Nick Vamivakas A 2015 Multi-dimensional single-spin nano-optomechanics with a levitated nanodiamond *Nat. Photonics* **9** 653–7

[24] Millen J, Deesuwan T, Barker P and Anders J 2014 Nanoscale temperature measurements using non-equilibrium Brownian dynamics of a levitated nanosphere *Nat. Nanotechnol.* **9** 425–9

[25] Gieseler J and Millen J 2018 Levitated nanoparticles for microscopic thermodynamics: a review *Entropy* **20** 326

[26] Li T, Kheifets S, Medellin D and Raizen M G 2010 Measurement of the instantaneous velocity of a Brownian particle *Science* **328** 1673–5

[27] Yin Z-Q, Geraci A A and Li T 2013 Optomechanics of levitated dielectric particles *Int. J. Mod. Phys. B* **27** 1330018

[28] Millen J, Monteiro T S, Pettit R and Vamivakas A N 2020 Optomechanics with levitated particles *Rep. Prog. Phys.* **83** 026401

[29] Ranjit G, Cunningham M, Casey K and Geraci A A 2016 Zeptonewton force sensing with nanospheres in an optical lattice *Phys. Rev. A* **93** 053801

[30] Kuhn S, Kosloff A, Stickler B A, Patolsky F, Hornberger K, Arndt M and Millen J 2017 Full rotational control of levitated silicon nanorods *Optica* **4** 356–60

[31] Reimann R, Doderer M, Hebestreit E, Diehl R, Frimmer M, Windey D, Tebbenjohanns F and Novotny L 2018 GHz rotation of an optically trapped nanoparticle in vacuum *Phys. Rev. Lett.* **121** 033602

[32] Ahn J, Xu Z, Bang J, Deng Y-H, Hoang T M, Han Q, Ma R-M and Li T 2018 Optically levitated nanodumbbell torsion balance and GHz nanomechanical rotor *Phys. Rev. Lett.* **121** 033603

[33] Bonin K D, Kourmanov B and Walker T G 2002 Light torque nanocontrol, nanomotors and nanorockers *Opt. Express* **10** 984–9

[34] MacDonald M P, Spalding G C and Dholakia K 2003 Microfluidic sorting in an optical lattice *Nature* **426** 421–4

[35] Ladavac K, Kasza K and Grier D G 2004 Sorting mesoscopic objects with periodic potential landscapes: optical fractionation *Phys. Rev. E* **70** 010901

[36] Paterson L, Papagiakoumou E, Milne G, Garcés-Chávez V, Tatarkova S A, Sibbett W, Gunn-Moore F J, Bryant P E, Riches A C and Dholakia K 2005 Light-induced cell separation in a tailored optical landscape *Appl. Phys. Lett.* **87** 123901

[37] Čižmár T, Šiler M, Šerý M, Zemánek P, Garcés-Chávez V and Dholakia K 2006 Optical sorting and detection of submicrometer objects in a motional standing wave *Phys. Rev. B* **74** 035105

[38] Rodrigo P J, Perch-Nielsen I R and Glückstad J 2006 Three-dimensional forces in GPC-based counterpropagating-beam traps *Opt. Express* **14** 5812–22

[39] Dholakia K, MacDonald M P, Zemánek P and Čižmár T 2007 Cellular and colloidal separation using optical forces *Methods in Cell Biology* ed M Berns and K Greulich (Cambridge, MA: Academic) pp. 467–95

[40] Eriksson E, Scrimgeour J, Granéli A, Ramser K, Wellander R, Enger J, Hanstorp D and Goksör M 2007 Optical manipulation and microfluidics for studies of single cell dynamics *J. Opt. A: Pure Appl. Opt.* **9** S113–21

[41] Atajanov A, Zhbanov A and Yang S 2018 Sorting and manipulation of biological cells and the prospects for using optical forces *Micro Nano Syst. Lett.* **6** 2

IOP Publishing

Optical Nanomanipulation (Second Edition)

David L Andrews and David S Bradshaw

Chapter 7

Biological applications of optical forces

A subject area that has been revolutionised, due to the adoption of optical trapping procedures, is the study and applications of nanoscale forces in biology. Among the key advantages of such methods are their capacity for non-invasive manipulation of target cells and biomolecules, achieved without photodamage. Especially interesting opportunities include the capacity for analysis of biomolecular motors and determination of the mechanical properties, such as elasticity, of DNA molecules. In addition, using a technique similar to optical fractionation, automated cell sorting via optical tweezers has been demonstrated, obviating the fluorescent markers required in conventional methods. While the first *in vitro* experiments were performed in the mid-80s, *in vivo* studies have become increasingly common in the last decade, which represent a new cutting edge in optical nanomanipulation.

7.1 Optical trapping of microbiological particles

For most biological applications, optical tweezer set-ups are far superior to the levitation techniques described in the previous chapter. Although originally designed as a trap for atoms, the optical tweezing of microbiological particles quickly gained favour. It is easy to understand why: the drawback of much greater mass can be compensated by suspension in a liquid (usually aqueous) medium, while the much larger surface area affords a far more effective response to the laser light. Moreover, the relatively slow rates of particle diffusion for sub-micron (compared to atomic scale) particles in the liquid provide for more easily controlled manipulation. The earliest tweezers could trap individual tobacco mosaic viruses (a rod-like protein) and *Escherichia coli* bacteria [1]; by avoiding any optical damage, even the process of cell reproduction of the latter can be observed within the trap [2].

To circumvent any photodestruction of biological particles over time, a near-infrared wavelength range of 750–1100 nm (conveniently produced by a titanium: sapphire laser) is usually chosen for the trapping beam rather than visible light—most biological samples are transparent in this range, so that damage due to light

doi:10.1088/978-0-7503-4191-2ch7

absorption is precluded. However, cell impairment through multiphoton absorption of near-infrared light has been reported [3]. Today, optical tweezing is no longer restricted to the trapping of single microbiological particles. Conventional single-beam tweezing methods have been replaced by multiple traps [4]—for example, red blood cells can now be trapped in an optical array (figure 7.1)—and techniques that use laser beams with exotic forms of wavefront structure [5]. Optical trapping arrays are discussed in the next chapter, while manipulation using structured light beams is detailed in chapters 9 and 10.

Within the last decade the trapping of biomolecular substances *in vivo*, such as red blood cells in mice [6–8], has significantly advanced and now represents a rapidly growing field [9]. One especially interesting study centres on embryonic zebrafish, in which direct control over the delicate motions of otoliths (inner ear, balance-detecting biostructures) is achieved through the application of optical trapping methods. Studies such as this provide unprecedented insights into how the processing of neural signals from these biostructures, whose role is to enable a creature to ascertain its disposition in conventional gravity conditions, produces appropriately adaptive movements of the eyes and tail [10, 11]. Detail on this application of bio-optical manipulation is shown in figure 7.2.

In the fascinating area of *neurophotonics*, light can play many different roles, including novel forms of optical manipulation. It is now well-established, for example, that laser light can affect the direction of neuron development and growth, though the detailed mechanisms are not fully understood [12, 13]. The very notion that such processes could be responsive to light was itself a surprise, representing a major discovery [14, 15].

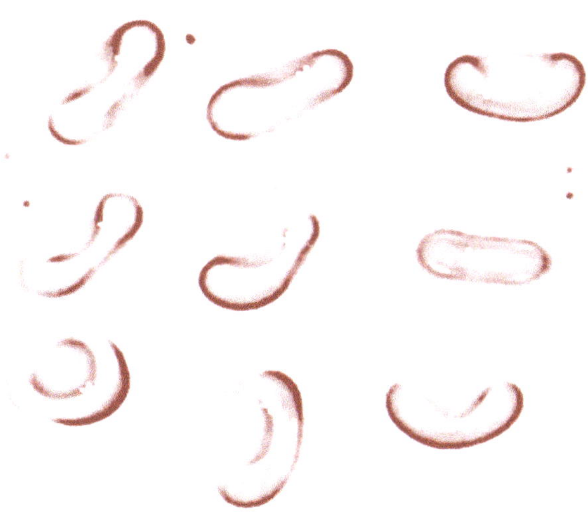

Figure 7.1. Multiplexed optical trapping of human red blood cells in a 3 × 3 array produced with an acousto-optic modulator. Each cell is approximately 6–8 μm in diameter. The laser operates at 1064 nm with 900 mW output power. Image courtesy of Kishan Dholakia, University of St Andrews.

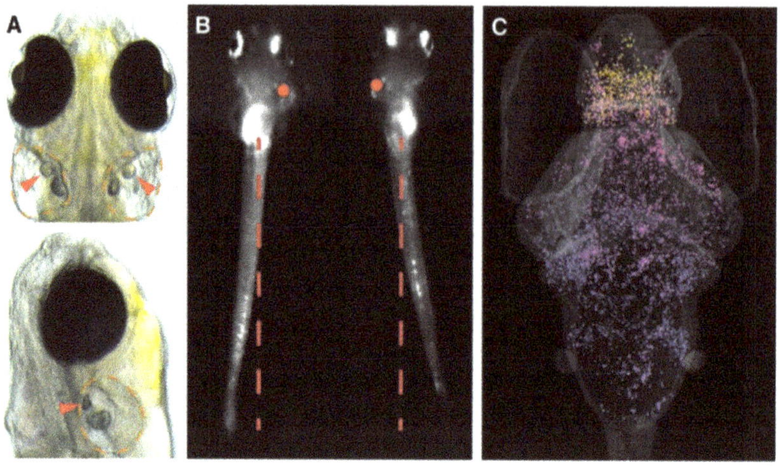

Figure 7.2. Manipulating the inner ear of a zebrafish with an optical trap. (A) Top view (upper photograph) and side view (lower) of a zebrafish. The inner ear is circled in dashed orange and the acceleration-sensing otolith is identified with a red arrowhead. (B) The tail bends in response to single otolith exterior trapping (red dot). The original position of the tail is shown by the dashed red line. (C) Distribution of neuronal intensity profiles in the zebrafish's brain. Reprinted by permission from De Gruyter: *Nanophotonics*, from reference [11], copyright 2019. Images courtesy of Gilles Vanwalleghem and Nick Valmas, University of Queensland.

7.2 Force measurements of individual biomolecules

Another significant application of optical tweezers is the study of biomolecular motors. (This differs from the appealing but rather fanciful notion that light-driven and guided nanomotors might find extensive use for biomedical applications *in vivo*, for example to non-intrusively facilitate complicated surgery within the body [16]. Interestingly, however, directed growth of suitably prepared nerve fibres has been demonstrated using light activated micromotors [17].) For example, optical force measurements represent a unique capability for examining the biophysics of some viral infection processes. The dynamic behaviour of viruses, which propagate genetic information between cells, can be determined by analysis of their force-extension and torque-rotation measurements as a function of time: the methods are similar to those applied in DNA studies, discussed below. By such means, mechanistic models can be quantitatively verified. A major triumph for this technique is the discovery that bacteriophages (viruses that infect and replicate within bacteria) are extremely powerful nanomotors that operate in a highly coordinated manner [18]. Using optical tweezer procedures to measure the packaging of DNA into single viral capsids [19, 20] reveals forces that are typically on the piconewton scale. Protecting and holding the DNA tightly within these structures generates internal pressures as high as 50 atm or more.

Arguably the foremost accomplishment of optical tweezers, in connection to biomolecular motors, is the discovery of mechanoenzyme interactions with the micro-tubules or actin filaments within cells to produce forces that are responsible for movement: examples include cell motility and muscle action. Before the advent of optical tweezers, the understanding of biomolecular processes was limited by the fact that only bulk biological samples could be examined. *In vitro* tweezing enabled unprecedented insight into the interactions of these biological motors (and other biopolymers) at the molecular level. For instance, Block *et al* observed the motor protein known as kinesin traverse a microtubule in a sequence of 8 nm steps [21] and piconewton-sized forces were recorded for kinesin motion [22]; an illustration of a kinesin molecule progressing along a microtubule is shown in figure 7.3. This study uses the 'handles' technique, in which the biomolecule of interest is attached to a small dielectric microsphere that is optically manipulated.

The method is also useful for applying a mechanical force across a biomolecule, as illustrated by figure 7.4, so that its properties (such as elasticity) can be found. For example, on attaching straddle beads (the 'handles') to each end of a DNA strand, the relationship between the applied force and the extension of the DNA molecule can be revealed [23–25] and the molecular process of its folding can be measured [26]—this is essential to know since the flexibility and winding of DNA play a crucial role in cellular functionality. Beyond simple stretching experiments, the fashioning of a knot into a single DNA strand (with a handle at each end) has also been achieved using optical tweezers [27]. The creation of more complex knots, and the analysis of their behaviour, could lead to a better understanding of how DNA remains knot-free inside a cell [28].

Other common bio-optical trapping geometries, outside the dual-trap geometry described above, are those in which one end of the biological structure is attached to the trapped bead and the other bound to a surface [29], or to a second bead that is held in position by suction at the tip of a micropipette [30]. More advanced tweezing methods are presented in the primer articles of references [31, 32]. One such

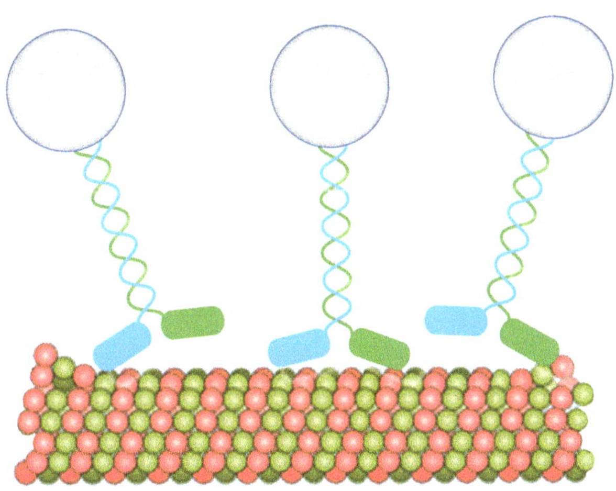

Figure 7.3. Molecular motor kinesin with a silica bead attached, depicted in three different stages of progression along the surface of a microtubule. Using optical tweezers to arrest the bead enables the motive force to be determined.

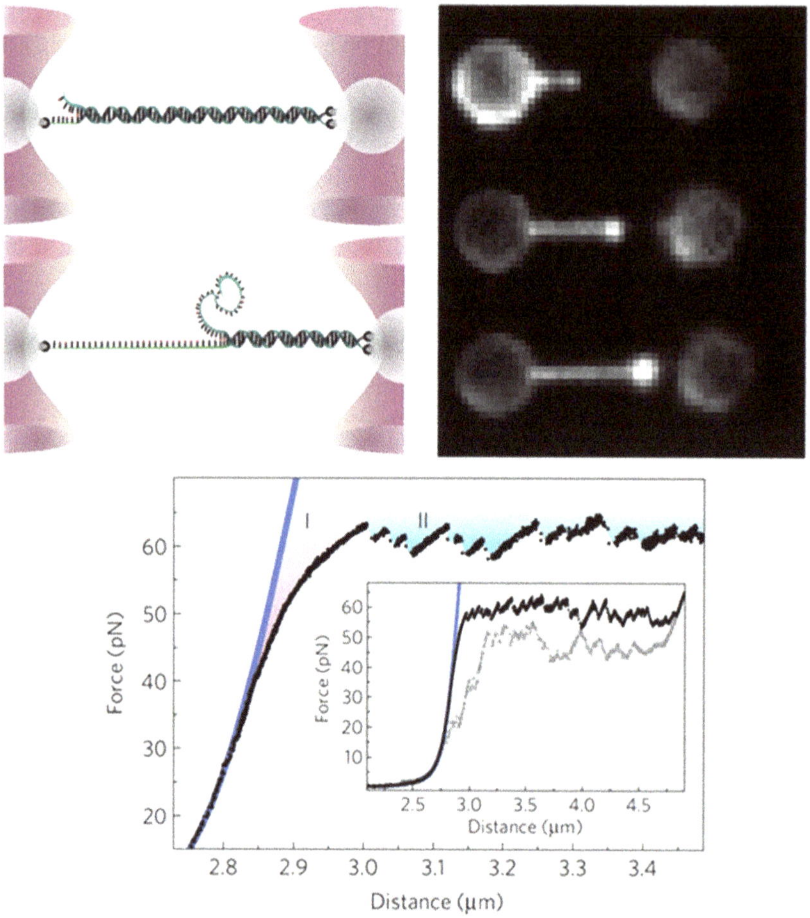

Figure 7.4. Force-induced strand unpeeling of DNA studied with fluorescence microscopy and force spectroscopy. (Top left) Schematic depictions of DNA undergoing unpeeling on extension, where three ends of the DNA strands are connected to the optically trapped microspheres. (Top right) Three fluorescence images of partially overstretched DNA molecules. Only the left part of the DNA construct is fluorescent, indicating unpeeling occurs only from the left side of the DNA. (Bottom) Force-extension measurement of the DNA construct. Region I: a strong deviation from the extensible model (blue line) is evident. Region II: an overstretching transition with pronounced stepwise increases in length is discerned. Inset: extended view of the same curve (black) together with the reverse process, in which strand reannealing occurs on relaxation (grey). Reprinted by permission from Springer Nature: Nature. *Nature Physics*, from reference [25], copyright 2011.

technique, known as *angular optical tweezers* (figure 7.5), is used to control and detect the torsional and rotational properties of individual biomolecules via the input of circularly polarized light. This is achieved by use of an optically trapped quartz cylinder, on which a torque can be exerted and measured, that is attached to the biomolecule of interest [33, 34]. Innovative optical tweezers techniques also include *fleezers*, in which the trapping radiation plays an additional, detection role by

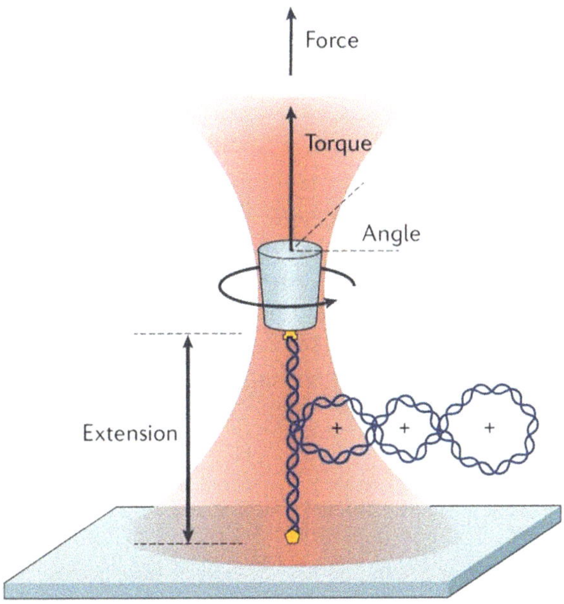

Figure 7.5. Depiction of the operational geometry of angular optical tweezers. A quartz cylinder is optically trapped in a beam (red concave cylinder), which is attached to a DNA molecule (blue double-helical lines) in this example. This technique is used to investigate the torsional properties of the DNA by applying a twist to it. Alongside the simultaneous exertion and measure of the torque, this method can also control the rotation and supercoiling of the DNA and measure the displacement and applied force on the trapped cylinder. Reprinted by permission from Springer Nature: Nature. *Nature Reviews Methods Primers*, from reference [31], copyright 2021.

instigating fluorescence and resonance energy transfer within the trapped species [35], and ones based on Raman spectroscopy. In the latter, molecular-level speciation of the trapped particles is achieved through the identification of frequency shifts in the spectrum of deflected light. Such methods are used for the examination of cells [36–38], nanoscale extracellular vesicles [39] and nanoplastics in sea water [40].

7.3 Cell sorting

The deployment of laser light for both trapping and speciation provides an especially useful means of achieving automated cell sorting. The latter requires three main steps: identification of the cell to be sorted, separation of the relevant cells and positioning of the cells into a designated reservoir. A common industrial standard for sorting a heterogeneous mixture of biological cells in flow cytometry is a set-up known as the *fluorescence-activated cell sorter* [41–43]. In this system, the cells (often attached to a fluorescent marker) are suspended in a rapidly flowing stream of liquid. While traversing the apparatus, the cells are first excited by a laser beam, and the detection of the subsequent characteristic fluorescence is used to identify the cell type. Depending on the latter, the cells are deflected into a specific direction and then collected into the appropriate reservoir. The main advantage of cell sorting by optical manipulation is that it is a completely sterile, non-contact procedure.

Notwithstanding the success and widespread adoption of fluorescence techniques, optical trapping separation methods have emerged that do not require fluorescent markers [44–48]. Here, cells are identified by the differences in their optomechanical response to the applied optical force, which offers a number of other key advantages [44]. Rather than sorting on the bulk scale, these passive cell sorters are capable of operating on a single cell basis, so that the sample size can be significantly reduced. This type of sorting method has a high positional accuracy, since the technique acts on each cell rather than a volume of fluid that contains the cell. It is interesting that, in its conception, cell sorting via optical trapping is similar to the optical fractionation procedure of the previous chapter.

References

[1] Ashkin A and Dziedzic J M 1987 Optical trapping and manipulation of viruses and bacteria *Science* **235** 1517–20

[2] Ashkin A, Dziedzic J M and Yamane T 1987 Optical trapping and manipulation of single cells using infrared laser beams *Nature* **330** 769–71

[3] König K, Liang H, Berns M W and Tromberg B J 1996 Cell damage in near-infrared multimode optical traps as a result of multiphoton absorption *Opt. Lett.* **21** 1090–2

[4] Chiou P Y, Ohta A T and Wu M C 2005 Massively parallel manipulation of single cells and microparticles using optical images *Nature* **436** 370–2

[5] Arlt J, Garces-Chavez V, Sibbett W and Dholakia K 2001 Optical micromanipulation using a Bessel light beam *Opt. Commun.* **197** 239–45

[6] Zhong M-C, Wei X-B, Zhou J-H, Wang Z-Q and Li Y-M 2013 Trapping red blood cells in living animals using optical tweezers *Nat. Commun.* **4** 1768

[7] Zhu R, Avsievich T, Popov A and Meglinski I 2020 Optical tweezers in studies of red blood cells *Cells* **9** 545

[8] Avsievich T, Zhu R, Popov A, Bykov A and Meglinski I 2020 The advancement of blood cell research by optical tweezers *Rev. Phys.* **5** 100043

[9] Xin H, Li Y, Liu Y-C, Zhang Y, Xiao Y-F and Li B 2020 Optical forces: from fundamental to biological applications *Adv. Mater.* **32** 2001994

[10] Favre-Bulle I A, Stilgoe A B, Rubinsztein-Dunlop H and Scott E K 2017 Optical trapping of otoliths drives vestibular behaviours in larval zebrafish *Nat. Commun.* **8** 630

[11] Favre-Bulle I A, Stilgoe A B, Scott E K and Rubinsztein-Dunlop H 2019 Optical trapping *in vivo*: theory, practice, and applications *Nanophoton.* **8** 1023–40

[12] Jesse F, Miao Z, Zhao L, Chen Y and Lv Y Y 2013 Optical fiber light source directs neurite growth *Biomed. Opt. Express* **4** 614–8

[13] Thompson A C, Stoddart P R and Jansen E D 2014 Optical stimulation of neurons *Curr. Mol. Imag.* **3** 162–77

[14] Wollman Y, Rochkind S and Simantov R 1996 Low power laser irradiation enhances migration and neurite sprouting of cultured rat embryonal brain cells *Neurol. Res.* **18** 467–70

[15] Ehrlicher A, Betz T, Stuhrmann B, Koch D, Milner V, Raizen M G and Käs J 2002 Guiding neuronal growth with light *Proc. Natl Acad. Sci. USA* **99** 16024

[16] Wang J, Xiong Z, Zheng J, Zhan X and Tang J 2018 Light-driven micro/nanomotor for promising biomedical tools: principle, challenge, and prospect *Acc. Chem. Res.* **51** 1957–65
[17] Wu T, Nieminen T A, Mohanty S, Miotke J, Meyer R L, Rubinsztein-Dunlop H and Berns M W 2012 A photon-driven micromotor can direct nerve fibre growth *Nat. Photonics* **6** 62–7
[18] Arias-Gonzalez J R and Mateu M G 2013 Optical tweezers to study viruses *Structure and Physics of Viruses: An Integrated Textbook* (Dordrecht: Springer) pp. 273–304
[19] Rickgauer J P, Fuller D N, Grimes S, Jardine P J, Anderson D L and Smith D E 2008 Portal motor velocity and internal force resisting viral DNA packaging in bacteriophage ϕ29 *Biophys. J.* **94** 159–67
[20] Keller N, delToro D J and Smith D E 2018 Single-molecule measurements of motor-driven viral DNA packaging in bacteriophages ϕ29, Λ, and T4 with optical tweezers *Molecular Motors: Methods and Protocols,* ed C Lavelle (New York: Springer) pp. 393–422
[21] Svoboda K, Schmidt C F, Schnapp B J and Block S M 1993 Direct observation of kinesin stepping by optical trapping interferometry *Nature* **365** 721–7
[22] Svoboda K and Block S M 1994 Force and velocity measured for single kinesin molecules *Cell* **77** 773–84
[23] Wang M D, Yin H, Landick R, Gelles J and Block S M 1997 Stretching DNA with optical tweezers *Biophys. J.* **72** 1335–46
[24] Bustamante C, Bryant Z and Smith S B 2003 Ten years of tension: single-molecule DNA mechanics *Nature* **421** 423–7
[25] Gross P, Laurens N, Oddershede L B, Bockelmann U, Peterman E J G and Wuite G J L 2011 Quantifying how DNA stretches, melts and changes twist under tension *Nat. Phys.* **7** 731–6
[26] Neupane K, Foster D A N, Dee D R, Yu H, Wang F and Woodside M T 2016 Direct observation of transition paths during the folding of proteins and nucleic acids *Science* **352** 239–42
[27] Arai Y, Yasuda R, Akashi K I, Harada Y, Miyata H, Kinosita K and Itoh H 1999 Tying a molecular knot with optical tweezers *Nature* **399** 446–8
[28] Bao X R, Lee H J and Quake S R 2003 Behavior of complex knots in single DNA molecules *Phys. Rev. Lett.* **91** 265506
[29] Yin H, Wang M D, Svoboda K, Landick R, Block S M and Gelles J 1995 Transcription against an applied force *Science* **270** 1653–7
[30] Smith S B, Cui Y and Bustamante C 1996 Overstretching B-DNA: the elastic response of individual double-stranded and single-stranded DNA molecules *Science* **271** 795–99
[31] Bustamante C J, Chemla Y R, Liu S and Wang M D 2021 Optical tweezers in single-molecule biophysics *Nat. Rev. Methods Primers* **1** 25
[32] Perkins T T 2009 Optical traps for single molecule biophysics: a primer *Laser Photon. Rev.* **3** 203–20
[33] Deufel C, Forth S, Simmons C R, Dejgosha S and Wang M D 2007 Nanofabricated quartz cylinders for angular trapping: DNA supercoiling torque detection *Nat. Methods* **4** 223–5
[34] Le T T, Gao X, Park S H, Lee J, Inman J T, Lee J H, Killian J L, Badman R P, Berger J M and Wang M D 2019 Synergistic coordination of chromatin torsional mechanics and topoisomerase activity *Cell* **179** 619–31

[35] Whitley K D, Comstock M J, Chemla Y R and Gennerich B 2017 High-resolution 'fleezers': dual-trap optical tweezers combined with single-molecule fluorescence detection *Optical Tweezers. Methods Molecular Biology* (New York: Harvard University Press) pp. 183–256

[36] De Luca A C, Rusciano G, Ciancia R, Martinelli V, Pesce G, Rotoli B, Selvaggi L and Sasso A 2008 Spectroscopical and mechanical characterization of normal and thalassemic red blood cells by Raman tweezers *Opt. Express* **16** 7943–57

[37] Snook R D, Harvey T J, Correia Faria E and Gardner P 2008 Raman tweezers and their application to the study of singly trapped eukaryotic cells *Integr. Biol.* **1** 43–52

[38] Bankapur A, Zachariah E, Chidangil S, Valiathan M and Mathur D 2010 Raman tweezers spectroscopy of live, single red and white blood cells *PLoS One* **5** e10427

[39] Kruglik S G, Royo F, Guigner J-M, Palomo L, Seksek O, Turpin P-Y, Tatischeff I and Falcón-Pérez J M 2019 Raman tweezers microspectroscopy of circa 100 nm extracellular vesicles *Nanoscale* **11** 1661–79

[40] Gillibert R *et al* 2019 Raman tweezers for small microplastics and nanoplastics identification in seawater *Environ. Sci. Technol.* **53** 9003–13

[41] Herzenberg L A, Parks D, Sahaf B, Perez O, Roederer M and Herzenberg L A 2002 The history and future of the fluorescence activated cell sorter and flow cytometry: a view from Stanford *Clin. Chem.* **48** 1819–27

[42] Autebert J, Coudert B, Bidard F-C, Pierga J-Y, Descroix S, Malaquin L and Viovy J-L 2012 Microfluidic: an innovative tool for efficient cell sorting *Methods* **57** 297–307

[43] Shen Y, Yalikun Y and Tanaka Y 2019 Recent advances in microfluidic cell sorting systems *Sens. Actuators B Chem.* **282** 268–81

[44] Grover S C, Skirtach A G, Gauthier R C and Grover C P 2001 Automated single-cell sorting system based on optical trapping *J. Biomed. Opt.* **6** 14–22

[45] Jonáš A and Zemánek P 2008 Light at work: the use of optical forces for particle manipulation, sorting, and analysis *Electrophoresis* **29** 4813–51

[46] Zhang H and Liu K-K 2008 Optical tweezers for single cells *J. R. Soc. Interface* **5** 671–90

[47] Landenberger B, Höfemann H, Wadle S and Rohrbach A 2012 Microfluidic sorting of arbitrary cells with dynamic optical tweezers *Lab Chip* **12** 3177–83

[48] Ma B, Yao B, Peng F, Yan S, Lei M and Rupp R 2012 Optical sorting of particles by dual-channel line optical tweezers *J. Opt.* **14** 105702

Chapter 8

Optical trapping arrays

By a variety of techniques, it is possible to use the principles of laser trapping to create stable particle arrays, achieved through the optical generation of a lattice of potential energy local minima. Such applications represent an extremely useful and distinctive component in the modern field of optical manipulation. As is the case with simple laser traps, these trapping arrays can be designed for application to anything from individual atoms to micron-sized particles. In connection with ultracold atom confinement, optical lattice operations enable the exploitation of ultranarrow linewidth resonances, as for example in applications to the storage of quantum information. For nanometre-sized and larger particles randomly suspended in liquid media, the creation of optical traps can support a range of very different applications, including size-based particle sorting.

8.1 Overview

Optical trapping arrays—alternatively known as optical lattices or, more fancifully, 'artificial crystals of light'—essentially comprise a set of two-dimensional or three-dimensional traps, originally and most simply created by the interference of counter-propagating laser beams (figure 8.1) [1–4]. The resulting optical potential landscape exhibits a periodic pattern. In such a system, particles can be trapped within individual potential wells due to the gradient force (which is described in chapter 3). A variety of properties that characterise the lattice potential can be controlled in real-time—such as the periodicity [5]—through changes in the relative phase, polarisation, frequency and intensity of the laser beams. Historically, the interest in optical trapping arrays stems from the study of the laser cooling of atoms [6]. More recently, for particles at room temperatures, alternative methods for optical lattice production have been successfully implemented based on time-sharing, holographic and near-field coupling techniques.

doi:10.1088/978-0-7503-4191-2ch8

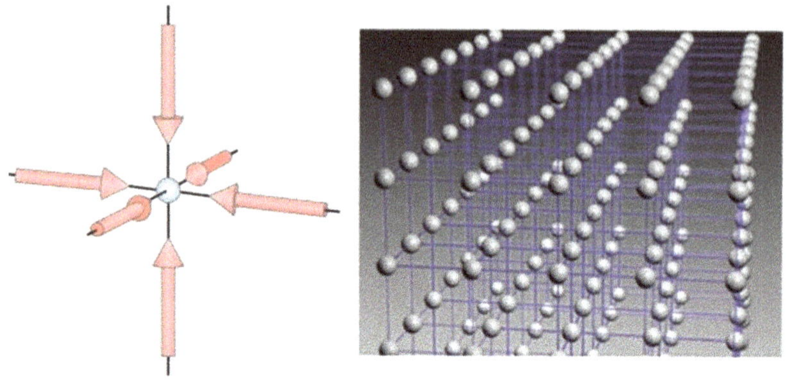

Figure 8.1. Three-dimensional optical lattice formed, at the co-located focus of three pairs of counter-propagating laser beams, through superimposition of the resultant standing waves in each dimension (left). The potential energy surface for such a lattice is approximated by a 3D simple cubic array of tightly confining harmonic oscillator potentials at each lattice site (right). Reprinted by permission from Springer Nature: Nature. *Nature Physics*, from reference [4], copyright 2005.

8.2 Ultracold atoms: optical lattices and quantum information

Optical lattices can be used to confine ultracold atoms [7–13]. The virtue of trapping atoms in a regular lattice, as opposed to the simpler case of a cloud—such as the Bose–Einstein condensates discussed in chapter 4—is that each of the isolated atoms experiences a very similar electronic environment, minimizing line broadening and decoherence. In dealing with the electronic properties of individual atoms, whose quantised energy levels are essentially discrete and relate to a narrow spectral linewidth, the polarisability is generally dominated by the first term within equation (3.8). This is especially so when the optical frequency of the input beam, ω, is similar to that of a specific electronic transition $0 \rightarrow s$, where s is the excited state in an assumed two-level system. Then, $E_{s0} \simeq \hbar\omega$, so that $\delta E = E_{s0} - \hbar\omega$ is small. To properly account for the optical response in such near-resonance conditions (i.e. the *dispersion* behaviour), it is necessary to introduce a damping term associated with the finite lifetime of excited states to the denominator of equation (3.6) [14]. The following expression is then found:

$$\alpha_{ij}(\omega) \simeq \frac{\mu_i^{0s}\mu_j^{s0}}{\delta E - \mathrm{i}\hbar\Omega_s}, \tag{8.1}$$

where Ω_s is the HWHM (half-width at half-maximum) of a free-atom resonance in the scattering profile. (The introduction of an imaginary damping term into the polarisability formula is essentially pragmatic. Its effect is to generate a Lorentzian form to the frequency dispersion near resonance, as a first approximation to a generic homogeneous lineshape.) From equation (8.1), it is clear that for red detuning, $\delta E > 0$, the polarisability is positive and the usual gradient force is attractive. Conversely, blue detuning produces a repulsive force, away from the most intense part of the beam. From the same expression, moreover, we see that the

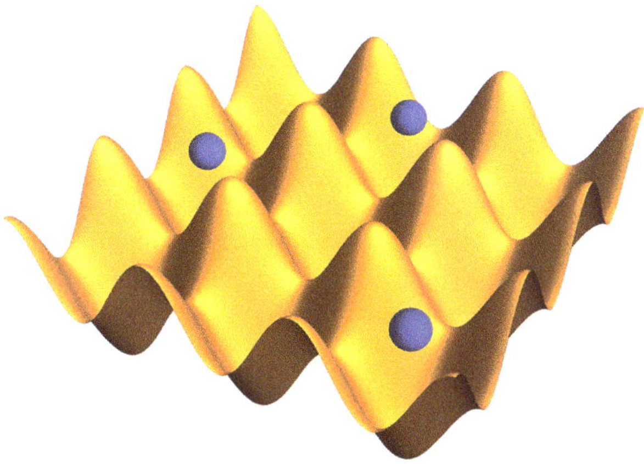

Figure 8.2. Depiction of an optical lattice potential (yellow surface), within which atoms (blue spheres) are trapped at local minima. This 'Atoms In Lattice' image has been obtained by the authors from the Wikimedia website, where it was made available by JPaget under a CC BY-SA 4.0 licence. It is included within this article on that basis. It is attributed to JPaget.

optical trapping of atoms involves their transition electric dipole moments. For cases beyond the dipole approximation, optical lattices based on electric quadrupole moments have been considered [15]. The motion of the trapped ultracold atoms is quantised and dependent on their vibrational motion within individual wells and the transfer of atoms between the potential minima (which may involve quantum tunnelling [16]); this is known as Bloch oscillation [17, 18].

The storing of ultracold atomic gases within optical lattices, as shown by figure 8.2, has led to numerous innovative possibilities [4, 19]. One of the most enticing prospects is for ultracold atoms trapped within optical arrays to be utilised as qubits for quantum information processing [20–27]. The possibility arises because ultracold atoms in their ground states couple extremely weakly to both the environment and each other, and as a result quantum decoherence (which leads irreversibly to a transformation of the qubits into classical bits) is suppressed. However, the weak form of interaction between the atoms is also a disadvantage, since a vital requisite for the register of a quantum computer is the phenomenon known as entanglement [28].

'On demand' methods to generate the required entanglement are based on either the application of an auxiliary laser beam to induce dipole–dipole interactions [29, 30] (using a method closely related to the optical binding techniques discussed in chapter 12) or 'cold controlled collisions' [31, 32]—a procedure that enhances local interactions by moving the potential wells closer together or by modifying the shape of the wells. The ability to activate interactions only when required is highly beneficial in overcoming decoherence. Experimental realisation of a quantum computer remains a major challenge. For instance, although some advances have been made [33, 34], a robust and reliable method is still required for addressing and reading the state of the qubits, complicated by atom separations that are generally smaller than an optical wavelength. Requirements such as the stabilisation of the local surroundings, which may encounter rogue electric and magnetic fields, are equally demanding.

Ultracold atoms within an optical lattice can also act as a basis for a high precision optical clock [35–37]. The extraordinarily accurate timekeeping in such a system is due to the narrow spectral linewidth of the trapped atoms, which is insensitive to external perturbations and effects such as the Doppler shift. These atomic clocks (often based on strontium) will neither gain nor lose a second over the estimated age of the Universe [38], and display higher accuracy than the caesium clocks used to determine International Atomic Time [39]—so much so that the SI unit of time, seconds, could be redefined in future. The reason for the improved precision is that optical lattice clocks measure light emitted in the visible range, rather than lower frequency microwaves, and also utilise a large number of trapped atoms so that signal errors are averaged out. Optical atomic clocks have many potential applications, which range from improved satellite navigation systems to ultrasensitive tests of fundamental physical theories.

8.3 Nanoparticles in suspension: techniques for optical lattice production

Time-sharing. When an optically trapped particle is dispersed within a liquid medium (for example, water), escape from the vicinity of the trap may be precluded by viscous forces even if the laser beam is removed for a short time. In such a scenario, any momentarily released particle will have a diffusion rate that conforms to Brownian motion, and the diffusion distance d of a particle for a beam removal time t is determined from;

$$d = \sqrt{\frac{2k_{\mathrm{B}}Tt}{3\pi\eta a}}, \tag{8.2}$$

where η denotes the viscosity of the medium. Hence, smaller particles have a capacity to migrate over larger diffusion distances, when the laser beam is absent. Consider, for example, a particle of 100 nm diameter immersed at room temperature, 298 K, in an aqueous medium ($\eta = 8.9 \times 10^{-4}$ Pa s). Calculation shows that particle translation by a full diameter can be expected in a timescale of approximately one millisecond. Ideally, the optical trap should, therefore, be removed for no longer than around 0.1 ms; indeed this is a typical value. Since the trapping laser can temporarily be moved to other positions (at great speed, for example, using computer-controlled deflectors), it is possible to simultaneously trap multiple particles by rapidly time-sharing the beam between each trapping site. Naturally, this time-sharing technique is limited by a maximum number of optical traps, which clearly depends on the size of the particles. The concept of a scanning laser trapping technique was demonstrated by Sasaki *et al* [40, 41]. Here, particles were aligned into a designated two-dimensional pattern by a single trapping laser that scans repeatedly across a specific route. Such systems, driven by mirror galvanometers, have been developed to achieve time-sharing rates around 1 kHz [42]—more than adequate for particles of micron size, but problematic for anything much smaller.

Currently, the most popular form of time-shared traps, offering tunable and highly precise beam-steering, are generated by acousto-optic deflectors (AODs)—as presented by Visscher *et al* [43]. The principle is based on a piezoelectric transducer generating an acoustic wave at MHz frequencies inside a transparent crystal, such as tellurium oxide. This is used to induce a standing wave modulation of refractive index along the path of a propagating light beam. The resulting diffraction can then be steered through an angle that is under the control of the signal frequency applied to the transducer. A pair of AODs perpendicularly aligned along the propagation axis provides for the generation of a number of optical traps, arbitrarily displaced over the focal plane. Computer-controlled AODs allow for a flexibility in the number and relative strengths of the optical traps, as well as control of the overall spatial pattern; adjustable time-sharing rates up to 10 kHz are achievable. While this technique is based on multi-particle manipulation within a plane, more elaborate optical schemes may generate additional layers [44]. Outside of systems based on AODs, electro-optic deflectors (EODs) can be used as an alternative scheme for modifying the refractive index gradient of the medium [45]. In such a case, which involves the application of an external electric field, the time-sharing rates are much faster than most systems that use AODs; however, a major downfall is that the attainable deflection angles are very small. Figure 8.3 offers a typical experimental set-up for the generation of multiple optical traps with AODs or EODs as the deflector.

Holography. Instead of time-sharing the trapping beam between numerous sites, it is ideally desirable to distribute the beam between each site simultaneously. A popular technique to achieve such an aim is based on holography [46–48]. Conventionally, the latter involves the splitting of a laser beam into two parts: one is reflected by an object before acting on a recording medium (the illumination beam), the other travels directly to the same photographic film (the reference beam). The interference pattern between the two beams is imprinted onto the recording medium, which is known as the hologram. On shining the original laser onto the developed film, the light is diffracted by the surface pattern of the hologram and an image of the object is generated.

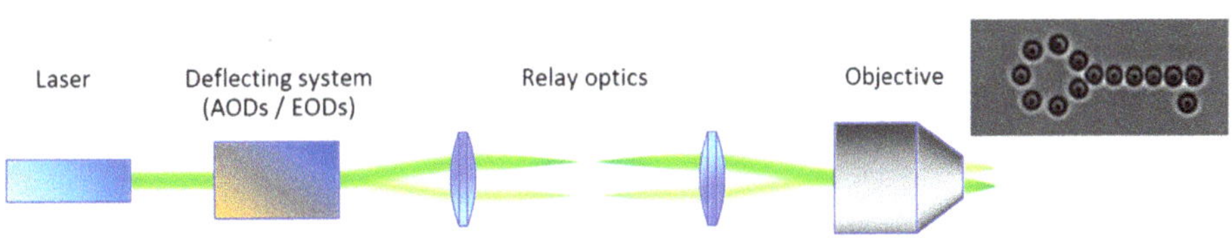

Figure 8.3. A typical experimental set-up for the generation of multiple optical traps using the time-sharing method. Although the laser beam is present at only one trapping position at any instant in time, fast switching between these positions enables up to several hundred time-shared optical traps to be sustained in one axial plane. The beam-steering element (acousto-optic deflectors, AODs, or electro-optic deflectors, EODs) and the objective back apertures are placed at conjugated planes of a relay telescope. Inset: typical optical lattice generated by this technique. Reproduced from reference [46]. Copyright IOP Publishing Ltd. All rights reserved.

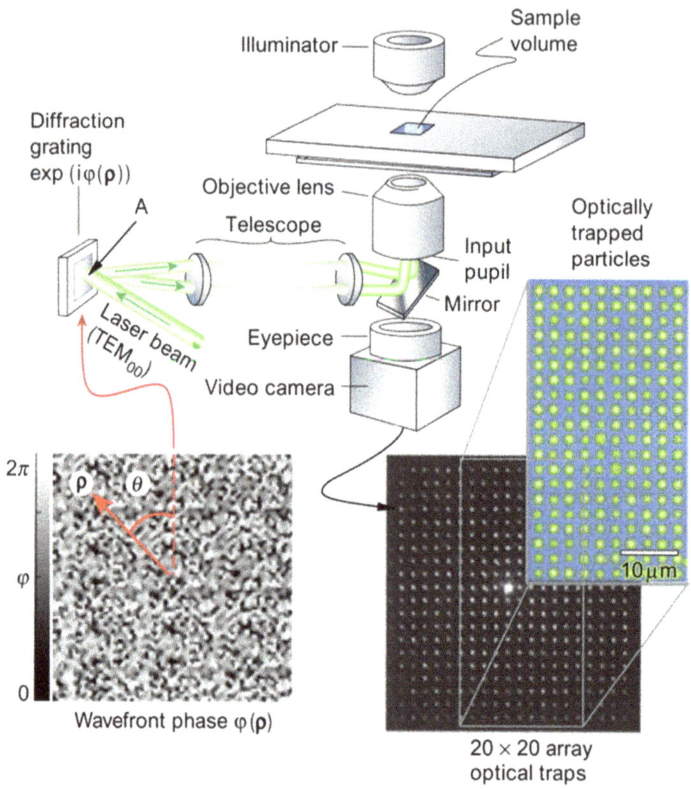

Figure 8.4. Simultaneous creation of multiple optical traps using computer-generated holograms. Projecting a collimated Gaussian (TEM_{00}) laser beam through the input pupil of a strongly converging lens, such as a microscope objective, creates a single optical trap. The telescope creates an image of the objective's input pupil, centred at point A. Multiple beams passing through point A pass into the objective lens to create multiple optical traps. A single Gaussian laser beam can be spilt into an arbitrary fan of beams, all emanating from point A, by an appropriate computer-designed diffraction grating. The example phase grating $\varphi(\boldsymbol{\rho})$ creates the 20 × 20 array of traps shown in the micrograph. These are shown trapping 800 nm diameter polystyrene spheres dispersed in water. Reprinted from reference [55], copyright 2002, with permission from Elsevier.

Although the concept of digital holography is implemented in holographic optical tweezers (HOTs), the technique does *not* require an actual object and no hologram is physically recorded. The optical trapping array is produced by diffracting laser light with a computer-generated hologram, with an appropriate algorithm fashioning the form of the array [49, 50]. Such a diffractive element (hologram), in its static form, is microfabricated by etching a surface pattern onto fused silica or a polymeric material [51, 52]. The alternative, dynamic diffractive element uses a computer-addressed spatial light modulator (SLM) to impose, in real time, a prescribed amount of phase shift on each liquid crystal pixel in a display [53–61]. Figure 8.4 schematically portrays a HOTs system that contains a dynamic diffractive element.

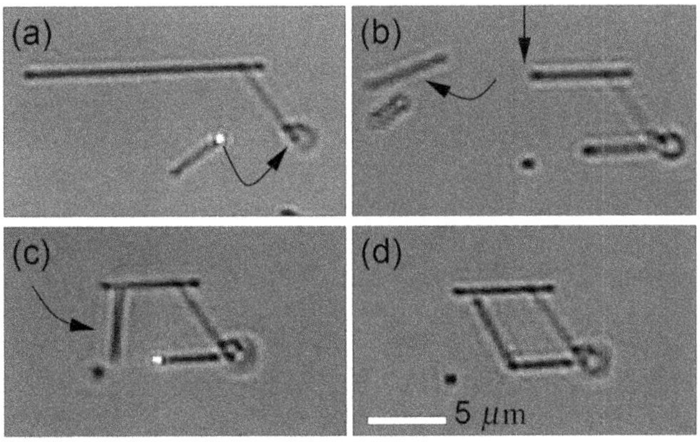

Figure 8.5. Assembly of a rhombus from semiconductor nanowires using holographic optical traps. (a) A nanowire is translated towards an existing structure, created earlier by trapping and fusing two nanowires. (b) The long nanowire is cut with a pulsed optical scalpel. (c) The resulting free-floating nanowire piece is brought towards the partially completed structure. (d) The structure is completed by fusing both ends of the nanowire piece. Reprinted with permission from reference [62], copyright The Optical Society.

Holographic optical tweezers have many advantages over the time-sharing schemes, including the capacity to create optical traps in three dimensions, shape each trap individually, arbitrarily distribute laser power between the traps and remove any optical aberrations. However, they require a complex computation process for large trapping arrays that is not required by other techniques, such as the generalised phase contrast method [63–66] (which has its own disadvantages, for example, the problem of highly focussing the trapping beams). Nanoscale applications of HOTs include the positioning and prising apart of carbon nanotube bundles [67], and the manipulation and assembly of nanowires [62, 68, 69]. The latter is depicted in figure 8.5, which shows the nanofabrication of a rhombus structure using holographic optical traps. Although optical lattices are usually created with Gaussian beams, systems based on Laguerre–Gaussian [55, 70] or Bessel [71, 72] laser light have been demonstrated, which produce more highly structured forms of trap. Complex beams such as these are discussed in detail in the following chapter.

Near-field coupling. Despite continued advancements in the creation of optical lattices, considerable challenges remain for separating nanoparticles larger than atoms, but smaller than 200 nm. Often such particles are synthesized in large quantities, by methods that produce a substantial distribution of sizes and, hence, a non-uniform response to trapping forces. Nonetheless, a promising novel laser configuration, based on an array of near-field coupled potential wells, has been designed to circumvent these restrictions [73]. Here, since the largest and smallest nanoparticles are, respectively, trapped by the deepest and shallowest potential wells, it is shown that the selective separation of nanoparticles, over the size range 100 to 500 nm, can be accomplished via a variation in the energy depth of the optical traps within the lattice. The latter is achieved by changing the power of the input laser. For example, with a judiciously chosen laser

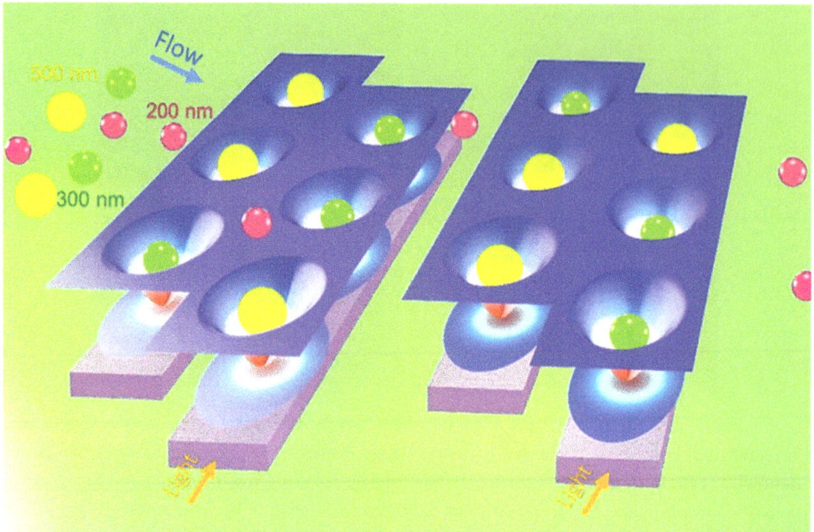

Figure 8.6. Depiction of a potential-energy array system for the selective trapping of nanoparticles with a distribution of sizes. In this configuration, the 300 and 500 nm particles are optically trapped in the potential wells, while the 200 nm particles are freely moving. Reprinted with permission from reference [73]. Copyright 2020 American Chemical Society.

power, nanoparticles with the sizes 300 and 500 nm can be optically trapped, while 200 nm particles remain free to move; this is illustrated in figure 8.6.

References

[1] Burns M M, Fournier J-M and Golovchenko J A 1990 Optical matter: crystallization and binding in intense optical fields *Science* **249** 749–54

[2] Chiou A E, Wang W, Sonek G J, Hong J and Berns M W 1997 Interferometric optical tweezers *Opt. Commun.* **133** 7–10

[3] Zemánek P, Jonáš A, Šrámek L and Liška M 1998 Optical trapping of Rayleigh particles using a Gaussian standing wave *Opt. Commun.* **151** 273–85

[4] Bloch I 2005 Ultracold quantum gases in optical lattices *Nat. Phys.* **1** 23–30

[5] Li T C, Kelkar H, Medellin D and Raizen M G 2008 Real-time control of the periodicity of a standing wave: an optical accordion *Opt. Express* **16** 5465–70

[6] Jessen P S and Deutsch I H 1996 Optical lattices *Advances in Atomic, Molecular, and Optical Physics* ed B Benjamin and W Herbert (Cambridge, MA: Academic) pp. 95–138

[7] Verkerk P, Lounis B, Salomon C, Cohen-Tannoudji C, Courtois J Y and Grynberg G 1992 Dynamics and spatial order of cold cesium atoms in a periodic optical potential *Phys. Rev. Lett.* **68** 3861–4

[8] Raithel G, Birkl G, Kastberg A, Phillips W D and Rolston S L 1997 Cooling and localization dynamics in optical lattices *Phys. Rev. Lett.* **78** 630–3

[9] Jaksch D, Bruder C, Cirac J I, Gardiner C W and Zoller P 1998 Cold bosonic atoms in optical lattices *Phys. Rev. Lett.* **81** 3108–11

[10] Guidoni L and Verkerk P 1999 Optical lattices: cold atoms ordered by light *J. Opt. B: Quantum Semiclass. Opt.* **1** R23–45

[11] Grynberg G and Robilliard C 2001 Cold atoms in dissipative optical lattices *Phys. Rep.* **355** 335–451
[12] Yukalov V I 2009 Cold bosons in optical lattices *Laser Phys.* **19** 1–110
[13] Krutitsky K V 2016 Ultracold bosons with short-range interaction in regular optical lattices *Phys. Rep.* **607** 1–101
[14] Andrews D L, Naguleswaran S and Stedman G E 1998 Phenomenological damping of nonlinear-optical response tensors *Phys. Rev. A* **57** 4925–9
[15] Moiseyev N, Šindelka M and Cederbaum L S 2007 Trapping of cold atoms in optical lattices by the quadrupole force *Phys. Lett. A* **362** 215–20
[16] Anderson B P and Kasevich M A 1998 Macroscopic quantum interference from atomic tunnel arrays *Science* **282** 1686–9
[17] Ben Dahan M, Peik E, Reichel J, Castin Y and Salomon C 1996 Bloch oscillations of atoms in an optical potential *Phys. Rev. Lett.* **76** 4508–11
[18] Kolovsky A R and Korsch H J 2003 Bloch oscillations of cold atoms in two-dimensional optical lattices *Phys. Rev. A* **67** 063601
[19] Lewenstein M, Sanpera A, Ahufinger V, Damski B, Sen A and Sen U 2007 Ultracold atomic gases in optical lattices: mimicking condensed matter physics and beyond *Adv. Phys.* **56** 243–379
[20] Deutsch I H, Brennen G K and Jessen P S 2000 Quantum computing with neutral atoms in an optical lattice *Fortschr. Phys.* **48** 925–43
[21] Monroe C 2002 Quantum information processing with atoms and photons *Nature* **416** 238–46
[22] Pachos J K and Knight P L 2003 Quantum computation with a one-dimensional optical lattice *Phys. Rev. Lett.* **91** 107902
[23] Porto J V, Rolston S, Laburthe Tolra B, Williams C J and Phillips W D 2003 Quantum information with neutral atoms as qubits *Philos. Trans. R. Soc. A* **361** 1417–27
[24] Jaksch D 2004 Optical lattices, ultracold atoms and quantum information processing *Contemp. Phys.* **45** 367–81
[25] Treutlein P *et al* 2006 Quantum information processing in optical lattices and magnetic microtraps *Fortschr. Phys.* **54** 702–18
[26] Bloch I, Dalibard J and Nascimbene S 2012 Quantum simulations with ultracold quantum gases *Nat. Phys.* **8** 267–76
[27] Saffman M 2016 Quantum computing with atomic qubits and Rydberg interactions: progress and challenges *J. Phys. B: At. Mol. Opt. Phys.* **49** 202001
[28] Amico L, Fazio R, Osterloh A and Vedral V 2008 Entanglement in many-body systems *Rev. Mod. Phys.* **80** 517–76
[29] Brennen G K, Caves C M, Jessen P S and Deutsch I H 1999 Quantum logic gates in optical lattices *Phys. Rev. Lett.* **82** 1060–3
[30] Brennen G K, Deutsch I H and Jessen P S 2000 Entangling dipole–dipole interactions for quantum logic with neutral atoms *Phys. Rev. A* **61** 062309
[31] Jaksch D, Briegel H J, Cirac J I, Gardiner C W and Zoller P 1999 Entanglement of atoms via cold controlled collisions *Phys. Rev. Lett.* **82** 1975–8
[32] Briegel H J, Calarco T, Jaksch D, Cirac J I and Zoller P 2000 Quantum computing with neutral atoms *J. Mod. Opt.* **47** 415–51

[33] Calarco T, Dorner U, Julienne P S, Williams C J and Zoller P 2004 Quantum computations with atoms in optical lattices: marker qubits and molecular interactions *Phys. Rev. A* **70** 012306

[34] Weitenberg C, Endres M, Sherson J F, Cheneau M, Schausz P, Fukuhara T, Bloch I and Kuhr S 2011 Single-spin addressing in an atomic Mott insulator *Nature* **471** 319–24

[35] Takamoto M, Hong F-L, Higashi R and Katori H 2005 An optical lattice clock *Nature* **435** 321–4

[36] Bloom B J, Nicholson T L, Williams J R, Campbell S L, Bishof M, Zhang X, Zhang W, Bromley S L and Ye J 2014 An optical lattice clock with accuracy and stability at the 10^{-18} level *Nature* **506** 71–5

[37] Ludlow A D, Boyd M M, Ye J, Peik E and Schmidt P O 2015 Optical atomic clocks *Rev. Mod. Phys.* **87** 637–701

[38] Derevianko A and Katori H 2011 Physics of optical lattice clocks *Rev. Mod. Phys.* **83** 331–47

[39] Le Targat R *et al* 2013 Experimental realization of an optical second with strontium lattice clocks *Nat. Commun.* **4** 2109

[40] Sasaki K, Koshioka M, Misawa H, Kitamura N and Masuhara H 1991 Pattern formation and flow control of fine particles by laser-scanning micromanipulation *Opt. Lett.* **16** 1463–5

[41] Sasaki K, Koshioka M, Misawa H, Kitamura N and Masuhara H 1992 Optical trapping of a metal particle and a water droplet by a scanning laser beam *Appl. Phys. Lett.* **60** 807–9

[42] Mio C, Gong T, Terray A and Marr D W M 2000 Design of a scanning laser optical trap for multiparticle manipulation *Rev. Sci. Instrum.* **71** 2196–200

[43] Visscher K, Gross S P and Block S M 1996 Construction of multiple-beam optical traps with nanometer-resolution position sensing *IEEE J. Sel. Top. Quant. Electron.* **2** 1066–76

[44] Vossen D L J, van der Horst A, Dogterom M and van Blaaderen A 2004 Optical tweezers and confocal microscopy for simultaneous three-dimensional manipulation and imaging in concentrated colloidal dispersions *Rev. Sci. Instrum.* **75** 2960–70

[45] Valentine M T, Guydosh N R, Gutiérrez-Medina B, Fehr A N, Andreasson J O and Block S M 2008 Precision steering of an optical trap by electro-optic deflection *Opt. Lett.* **33** 599–601

[46] Čižmár T D, Romero L C, Dholakia K and Andrews D L 2010 Multiple optical trapping and binding: new routes to self-assembly *J. Phys. B: At. Mol. Opt. Phys.* **43** 102001

[47] Grier D G 2003 A revolution in optical manipulation *Nature* **424** 810–6

[48] Woerdemann M, Alpmann C, Esseling M and Denz C 2013 Advanced optical trapping by complex beam shaping *Laser Photon. Rev.* **7** 839–54

[49] Kim H, Kim M, Lee W and Ahn J 2019 Gerchberg-Saxton algorithm for fast and efficient atom rearrangement in optical tweezer traps *Opt. Express* **27** 2184–96

[50] He M R, Liang Y S, Bianco P R, Wang Z J, Yun X, Cai Y N, Feng K and Lei M 2021 Trapping performance of holographic optical tweezers generated with different hologram algorithms *AIP Adv.* **11** 035130

[51] Dufresne E R and Grier D G 1998 Optical tweezer arrays and optical substrates created with diffractive optics *Rev. Sci. Instrum.* **69** 1974–7

[52] Dufresne E R, Spalding G C, Dearing M T, Sheets S A and Grier D G 2001 Computer-generated holographic optical tweezer arrays *Rev. Sci. Instrum.* **72** 1810–6

[53] Reicherter M, Haist T, Wagemann E U and Tiziani H J 1999 Optical particle trapping with computer-generated holograms written on a liquid-crystal display *Opt. Lett.* **24** 608–10

[54] Liesener J, Reicherter M, Haist T and Tiziani H J 2000 Multi-functional optical tweezers using computer-generated holograms *Opt. Commun.* **185** 77–82
[55] Curtis J E, Koss B A and Grier D G 2002 Dynamic holographic optical tweezers *Opt. Commun.* **207** 169–75
[56] Hossack W J, Theofanidou E, Crain J, Heggarty K and Birch M 2003 High-speed holographic optical tweezers using a ferroelectric liquid crystal microdisplay *Opt. Express* **11** 2053–9
[57] Melville H, Milne G F, Spalding G C, Sibbett W, Dholakia K and McGloin D 2003 Optical trapping of three-dimensional structures using dynamic holograms *Opt. Express* **11** 3562–7
[58] Leach J, Sinclair G, Jordan P, Courtial J, Padgett M J, Cooper J and Laczik Z J 2004 3D manipulation of particles into crystal structures using holographic optical tweezers *Opt. Express* **12** 220–6
[59] Sinclair G, Leach J, Jordan P, Gibson G, Yao E, Laczik Z J, Padgett M J and Courtial J 2004 Interactive application in holographic optical tweezers of a multi-plane Gerchberg-Saxton algorithm for three-dimensional light shaping *Opt. Express* **12** 1665–70
[60] Polin M, Ladavac K, Lee S-H, Roichman Y and Grier D G 2005 Optimized holographic optical traps *Opt. Express* **13** 5831–45
[61] Grier D G and Roichman Y 2006 Holographic optical trapping *Appl. Opt.* **45** 880–7
[62] Agarwal R, Ladavac K, Roichman Y, Yu G, Lieber C M and Grier D G 2005 Manipulation and assembly of nanowires with holographic optical traps *Opt. Express* **13** 8906–12
[63] Mogensen P C and Gluckstad J 2000 Dynamic away generation and pattern formation for optical tweezers *Opt. Commun.* **175** 75–81
[64] Eriksen R L, Daria V R and Glückstad J 2002 Fully dynamic multiple-beam optical tweezers *Opt. Express* **10** 597–602
[65] Rodrigo P J, Perch-Nielsen I R, Alonzo C A and Glückstad J 2006 GPC-based optical micromanipulation in 3D real-time using a single spatial light modulator *Opt. Express* **14** 13107–12
[66] Dam J S, Rodrigo P J, Perch-Nielsen I R, Alonzo C A and Glückstad J 2007 Computerized 'drag-and-drop' alignment of GPC-based optical micromanipulation system *Opt. Express* **15** 1923–31
[67] Plewa J, Tanner E, Mueth D M and Grier D G 2004 Processing carbon nanotubes with holographic optical tweezers *Opt. Express* **12** 1978–81
[68] Simpson S H and Hanna S 2010 Holographic optical trapping of microrods and nanowires *J. Opt. Soc. Am. A* **27** 1255–64
[69] Li J and Du G 2014 Manipulation and assembly of ZnO nanowires with single holographic optical tweezers system *Appl. Opt.* **53** 351–5
[70] Curtis J E and Grier D G 2003 Structure of optical vortices *Phys. Rev. Lett.* **90** 133901
[71] Čižmár T, Kollárová V, Tsampoula X, Gunn-Moore F, Sibbett W, Bouchal Z and Dholakia K 2008 Generation of multiple Bessel beams for a biophotonics workstation *Opt. Express* **16** 14024–35
[72] Lu Y, Jiang B, Lü S, Liu Y, Li S, Cao Z and Qi X 2016 Arrays of Gaussian vortex, Bessel and Airy beams by computer-generated hologram *Opt. Commun.* **363** 85–90
[73] Shi Y, Zhao H, Chin L K, Zhang Y, Yap P H, Ser W, Qiu C-W and Liu A Q 2020 Optical potential-well array for high-selectivity, massive trapping and sorting at nanoscale *Nano Lett.* **20** 5193–200

David L Andrews and David S Bradshaw

Chapter 9

Orbital angular momentum, optical vortices and torques

Beams of light may convey more than energy, linear momentum and spin angular momentum. When suitably structured, the propagation of light can also deliver orbital angular momentum (OAM) in a quantised form. Such structured light is commonly referred to as an *optical vortex* or *twisted light*, for which a line of zero intensity develops along the axis of propagation. Its key attribute is a phase factor of the form $\exp(\pm il\phi)$, where l denotes the topological charge or winding number and the angle ϕ denotes azimuthal position around the beam axis. Light engendered with OAM includes the widely used Laguerre–Gaussian modes, and the commonest forms of Bessel and Mathieu beams. Within the high-intensity rings of such structured light, particles may be trapped and circulated. The presence of an optical torque due to the wavefront 'twist' is the origin of the latter.

9.1 Orbital angular momentum

Twisted beams are usually produced by conversion of a simple Gaussian-type laser beam. Experimental methods for bestowing OAM are legion, and they include the passage of a beam through pitch-fork holograms [1–3], spiral phase plates [4], paired cylindrical lenses [5], q-plates [6], hyperbolic metamaterials [7], spatial light modulators [8–10] or optical fibres [11]. Among other possibilities, a method has been proposed for the direct generation of light endowed with OAM, based on light emission from an array of nanoparticles [12–14]. To understand the potential that the light then acquires for manipulating nanoscale material, we now return to the theory of optical angular momentum, introduced in chapter 2.

The total angular momentum can be written in operator form as a sum of spin and orbital angular momentum contributions, as in the following expression: $\hat{\mathbf{J}}_{\text{rad}} = \hat{\mathbf{S}}_{\text{rad}} + \hat{\mathbf{L}}_{\text{rad}}$ (although this separation is only possible by use of the paraxial approximation; non-paraxial light is not considered here, but is examined

doi:10.1088/978-0-7503-4191-2ch9

elsewhere [15–18]). Here, the spin angular momentum, $\hat{\mathbf{S}}_{\text{rad}}$, is given by equation (2.12) and the OAM operator, $\hat{\mathbf{L}}_{\text{rad}}$, is expressible as;

$$\hat{\mathbf{L}}_{\text{rad}} = \sum_{k,\eta,l,p} \hat{N}_{lp}^{(\eta)}(\mathbf{k})\, l\hbar\mathbf{u}_{\mathbf{k}}, \tag{9.1}$$

which is cast in terms of Laguerre–Gaussian modes with four degrees of freedom: a wave-vector of magnitude k, polarisation η, and integers l and p that designate the order and degree of the corresponding Laguerre polynomial. The latter symbol is the radial index that represents the number of radial nodes, $p + 1$, within the transverse field distribution of the beam. In passing, it is interesting to note that by casting the sum over η explicitly in terms of a basis comprising left- and right-handed circular polarisations, we arrive at [19];

$$\hat{\mathbf{L}}_{\text{rad}} = \sum_{k,l,p} \left\{\hat{N}_{lp}^{(L)}(\mathbf{k}) + \hat{N}_{lp}^{(R)}(\mathbf{k})\right\} l\hbar\mathbf{u}_{\mathbf{k}}, \tag{9.2}$$

which forms a neat counterpart to the spin angular momentum given by equation (2.12). The above expressions indicate that the orbital angular momentum has no connection with any spin helicity associated with polarisation [20]. Furthermore, we can determine that a free-field photon endowed with topological charge has an OAM magnitude of $l\hbar$. Generally, l signifies the number of distinct helical surfaces formed by the optical wavefront, as it winds around the propagation axis within the span of a wavelength (figure 9.1).

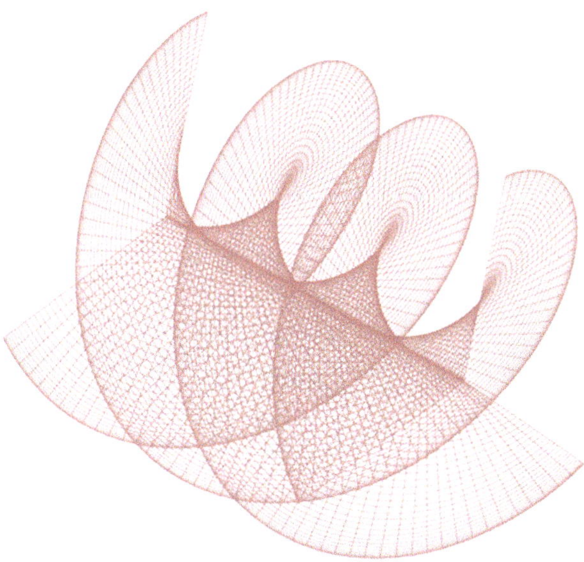

Figure 9.1. Wavefront surface for an optical vortex with a topological charge of three, over the span of three wavelengths (three intertwined helical surfaces).

9.2 Optical vortices

The theoretical basis for the concept of an optical vortex beam was first established in a series of works [21–23]. For Laguerre–Gaussian modes, the explicit formulations for the electric and magnetic field expansions—previously defined in equations (2.3) and (2.4) for plane waves—are written as follows [24];

$$\hat{\mathbf{E}}(\mathbf{r}) = \mathrm{i}\sum_{k,\eta,l,p}\left(\frac{\hbar ck}{2\varepsilon_0 V}\right)^{\frac{1}{2}} f_{l,p}(r)\left\{\mathbf{e}_{l,p}^{(\eta)}(\mathbf{k})\hat{a}_{l,p}^{(\eta)}(\mathbf{k})\mathrm{e}^{\mathrm{i}(\mathbf{k}\cdot\mathbf{r}+l\phi)} - \bar{\mathbf{e}}_{l,p}^{(\eta)}(\mathbf{k})\hat{a}^{\dagger}{}_{l,p}^{(\eta)}(\mathbf{k})\mathrm{e}^{-\mathrm{i}(\mathbf{k}\cdot\mathbf{r}+l\phi)}\right\}, \quad (9.3)$$

$$\hat{\mathbf{B}}(\mathbf{r}) = \mathrm{i}\sum_{k,\eta,l,p}\left(\frac{\hbar k}{2\varepsilon_0 cV}\right)^{\frac{1}{2}} f_{l,p}(r)\left\{\mathbf{b}_{l,p}^{(\eta)}(\mathbf{k})\hat{a}_{l,p}^{(\eta)}(\mathbf{k})\mathrm{e}^{\mathrm{i}(\mathbf{k}\cdot\mathbf{r}+l\phi)} - \bar{\mathbf{b}}_{l,p}^{(\eta)}(\mathbf{k})\hat{a}^{\dagger}{}_{l,p}^{(\eta)}(\mathbf{k})\mathrm{e}^{-\mathrm{i}(\mathbf{k}\cdot\mathbf{r}+l\phi)}\right\}, \quad (9.4)$$

using cylindrical coordinates that comprise off-axial radial distance r, axial position z and azimuthal angle ϕ. In these expressions, $f_{l,p}(r)$ is a normalised radial function, whose form depends on l only through its modulus. In consequence, all information on wavefront *handedness* is found within the phase factors, in which a positive or negative sign for l represents a helical wavefront structure of left- or right-handed helicity, respectively. Achieving this type of structure places constraints upon the form of the transverse field distributions. Typical examples of these distributions for Laguerre–Gaussian modes are presented in figure 9.2, which specifically depicts the associated intensity and phase structure. Figure 9.3(a) gives an impression of how the wave-vector, everywhere pointing along the normal to the wavefront surface, effectively twists around the beam axis. Laguerre–Gaussian

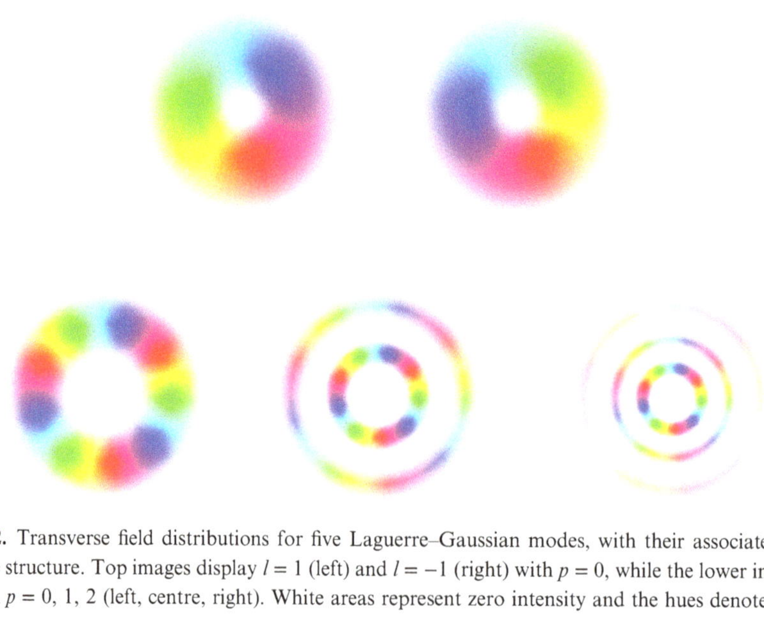

Figure 9.2. Transverse field distributions for five Laguerre–Gaussian modes, with their associated intensity and phase structure. Top images display $l = 1$ (left) and $l = -1$ (right) with $p = 0$, while the lower images show $l = 3$ with $p = 0, 1, 2$ (left, centre, right). White areas represent zero intensity and the hues denote the phase.

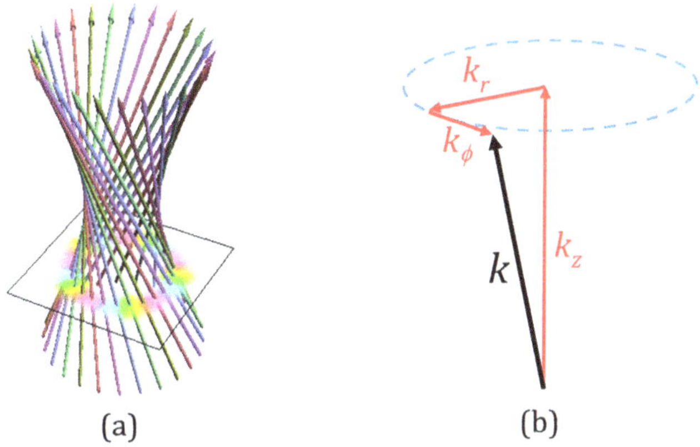

Figure 9.3. (a) Schematic representation for the focus of an $l = 5$ optical vortex mode. In the transverse plane, hue indicates the phase and brightness the intensity. The bundle of Poynting vectors, depicted as colour-coded arrows, rotates about the beam axis by one fifth of a circle over the space of one wavelength: (b) resolution of a local Poynting vector into axial (z), radial (r) and azimuthal (ϕ) components. Image (a) courtesy of Sonja Franke-Arnold, University of Glasgow.

beams with $p > 0$ are, in a sense, similar in nature to higher-order ($l > 0$) Bessel beams; however, the latter has non-diffracting and self-healing properties [25–28]. The electric and magnetic field expansions for the Bessel beam are comparable to equations (9.3) and (9.4), except that $f_{l,p}(r)$ is substituted with an lth-order Bessel function. Mathieu beams are essentially Bessel beams with elliptical (rather than circular) rings, with field expressions that are again correspondingly similar in form [29, 30].

9.3 Optical torque

Beams of light with various structures (both those with and without OAM) can trap and manipulate nanoparticles. Particles with conventional optical dispersion properties are typically attracted to the high intensity regions via the gradient force—i.e. the circular rings in the case of a Laguerre–Gaussian or a higher-order Bessel beam, as opposed to the centre of a regular Gaussian beam. However, as an additional feature, beams that contain OAM may also produce an optical torque. It should be emphasised here that such a torque is not the same as the one expressed in equation (3.9), which corresponds to the alignment of a non-spherical nanoparticle with the polarisation plane of the trapping beam due to the gradient force. The optical torque of present interest occurs even with spherical particles, producing orbital particle circulation around the beam, as indicated by the arrow within figure 9.4. This torque may arise due to the radiation force associated with *non-forward* Rayleigh scattering, in addition to the forces detailed in chapter 3 that occur if there is any light absorption. The radiation force will usually 'push' the

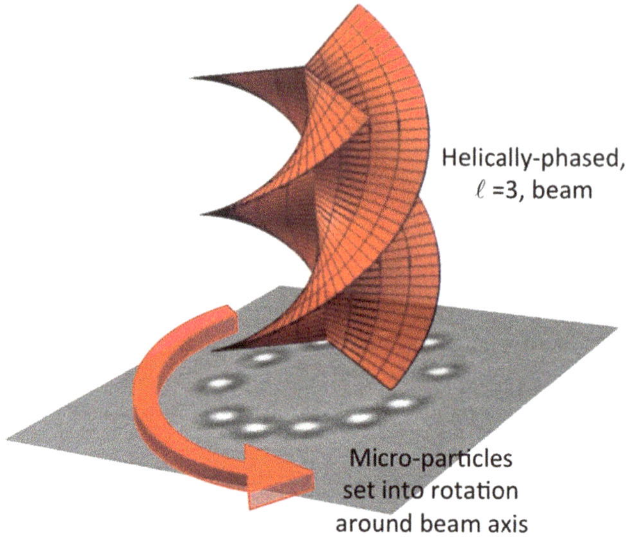

Figure 9.4. Helically phased laser beam incident on a set of microscopic particles. The latter are trapped within the high-intensity ring of the beam (the centre of an optical vortex beam has zero intensity, unlike Gaussian light) and orbit around the beam axis, as indicated by the red arrow. Image courtesy of Miles Padgett, University of Glasgow.

particles along the beam axis (explaining the need for a beam focus in the optical tweezers set-up, as observed in chapter 5). However, the inclusion of a helical wavefront means that a torque will also be applied to the nanoparticle. A quantum description of such a torque now follows.

The optical torque that originates from the helical wavefront can be determined from second-order perturbation theory. Here, in contrast to equation (3.6), the initial and final system states are *not* identical since the torque is based on a non-forward Rayleigh scattering mechanism. In such a process, the created and annihilated photons have identical energy but travel in different directions. Due to this disparity an additional radiation mode is assigned to the system states, so that $| I\rangle = | n, 0; 0\rangle$ and $| F\rangle = | (n-1), 1; 0\rangle$ for the present case. Here, the first and second elements in each state vector denote radiation modes relating to the annihilation and creation events, respectively, and the third element represents the particle state [31]. (Forward Rayleigh scattering, in contrast, entails photon annihilation and creation from and into the same radiation mode.) This subtle change in the initial and final system states, which contrasts to the forward Rayleigh scattering case of chapter 3, may seem innocuous but it facilitates a physically important differentiation between the radiation and gradient force. The optical torque is determined by multiplying the scattering rate (found from the Fermi rule) by the radius of the beam ring, r_b, and the orbital angular momentum per photon $l\hbar\mathbf{u}_{\mathbf{kz}}$, where $\mathbf{u}_{\mathbf{kz}}$ is the unit vector of $\mathbf{k}_\mathbf{z}$; the latter is illustrated in figure 9.3(b). Following a rotational average, assuming linearly polarised input and summing over all possible polarisations for the scattered light, the following expression in terms of polarisability components is derived for the observed torque [31];

$$\boldsymbol{\tau} = \frac{Ik^3 l r_b \mathbf{u}_{\mathbf{kz}}}{40\pi\varepsilon_0^2 c}\left(\alpha_{\lambda\lambda}\alpha_{\mu\mu} + 7\alpha_{\lambda\mu}\alpha_{\lambda\mu}\right), \tag{9.5}$$

where the Greek indices denote the particle frame of reference for the polarisability components (rather than the laboratory frame represented by the earlier Latin indices). We are reminded here that every particle of matter has a non-zero polarisability, though the magnitude of its components varies with wavelength through conventional dispersion relations [32].

In equation (9.5), the optical response of the particle is written in terms of two material parameters, each written using the implied summation convention encountered in chapter 3. Thus, $\alpha_{\lambda\lambda}\alpha_{\mu\mu}$ represents the square of the trace of the polarisability, tr($\boldsymbol{\alpha}$), and $\alpha_{\lambda\mu}\alpha_{\lambda\mu}$ is a sum of the squares of the nine polarisability components. Larger dielectric particles may be treated with electronic properties closer to those of a bulk material. It is appropriate, in such cases, to engage the linear optical susceptibility rather than the polarisability tensor; the connection between the two is explained in detail elsewhere [33]. In all situations, therefore, there are just two scalar parameters, which together effectively determine the particle's propensity to acquire angular momentum about the vortex beam axis.

The circular motion produced by action of an optical vortex will, of course, be subject to viscous forces acting as a drag, and these are simply determined by the standard viscosity of the host medium. The magnitude of the velocity is thus given by $v = \tau / c_d r_b$, where c_d is the drag coefficient. This orbiting behaviour was first theoretically predicted [34] and experimentally observed [35–39] with Laguerre–Gaussian light (sometimes known as an optical spanner or wrench)—such a system is now recognised as a foundation for the next generation of optical tweezers [40–44]. The technique has since been refined by a scheme involving the interference pattern of a Laguerre–Gaussian beam with a Gaussian laser, which circumvents problems such as the photodamage of particles due to heating. In such a scheme the number of 'arms' of the spiral pattern, produced by the beam interference and relating to the l value of the structured beam, can be tailored to fit the shape of the particle. Moreover, by changing the effective pathlength of one of the interfering beams, the speed of the orbit can be optically controlled [45]. More recently, a procedure based on two counter-propagating circularly polarised Laguerre–Gaussian beams has shown, via both simulations and experiment, that silicon nanowires can be spun and circulated simultaneously—this is due to a transfer of the spin and orbital angular momentum from the beams [46]. In a more exotic light structure involving interlocking vortex rings, the circulation of particles has also been demonstrated (figure 9.5).

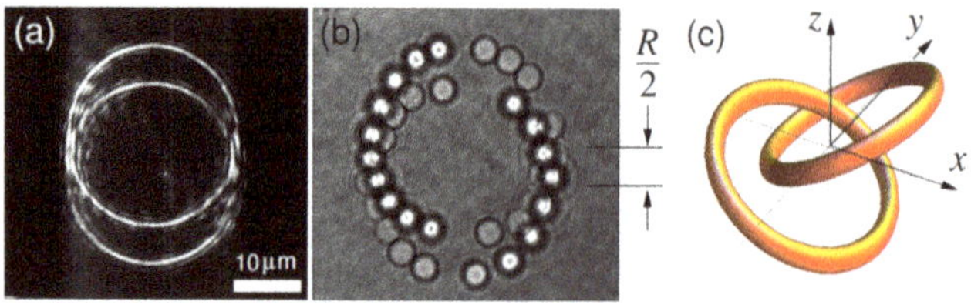

Figure 9.5. (a) Interlocked optical vortex traps; (b) particles circulating about each trap (note the depth profile); (c) depiction of the three-dimensional configuration. Image courtesy of David Grier, New York University.

References

[1] Bazhenov V Y, Vasnetsov M and Soskin M 1990 Laser beams with screw dislocations in their wavefronts *JETP Lett.* **52** 429–31

[2] Bazhenov V Y, Soskin M and Vasnetsov M 1992 Screw dislocations in light wavefronts *J. Mod. Opt.* **39** 985–90

[3] Mirhosseini M, Magana-Loaiza O S, Chen C, Rodenburg B, Malik M and Boyd R W 2013 Rapid generation of light beams carrying orbital angular momentum *Opt. Express* **21** 30196–203

[4] Beijersbergen M W, Coerwinkel R P C, Kristensen M and Woerdman J P 1994 Helical-wave-front laser-beams produced with a spiral phase plate *Opt. Commun.* **112** 321–7

[5] Beijersbergen M W, Allen L, Vanderveen H E L O and Woerdman J P 1993 Astigmatic laser mode converters and transfer of orbital angular-momentum *Opt. Commun.* **96** 123–32

[6] Marrucci L, Manzo C and Paparo D 2006 Optical spin-to-orbital angular momentum conversion in inhomogeneous anisotropic media *Phys. Rev. Lett.* **96** 163905

[7] Sun J, Zeng J and Litchinitser N M 2013 Twisting light with hyperbolic metamaterials *Opt. Express* **21** 14975–81

[8] Heckenberg N R, McDuff R, Smith C P and White A G 1992 Generation of optical-phase singularities by computer-generated holograms *Opt. Lett.* **17** 221–3

[9] Ostrovsky A S, Rickenstorff-Parrao C and Arrizón V 2013 Generation of the 'perfect' optical vortex using a liquid-crystal spatial light modulator *Opt. Lett.* **38** 534–6

[10] Forbes A, Dudley A and McLaren M 2016 Creation and detection of optical modes with spatial light modulators *Adv. Opt. Photon.* **8** 200–27

[11] Dong M, Yang Z, Chao Z, Hua L, Cong W, Han Z, Wending Z, Ting M and Jianlin Z 2021 Generation of polarization and phase singular beams in fibers and fiber lasers *Adv. Photon.* **3** 014002

[12] Coles M M, Williams M D, Saadi K, Bradshaw D S and Andrews D L 2013 Chiral nanoemitter array: a launchpad for optical vortices *Laser Photon. Rev.* **7** 1088–92

[13] Williams M D, Coles M M, Saadi K, Bradshaw D S and Andrews D L 2013 Optical vortex generation from molecular chromophore arrays *Phys. Rev. Lett.* **111** 153603

[14] Williams M D, Coles M M, Bradshaw D S and Andrews D L 2014 Direct generation of optical vortices *Phys. Rev. A* **89** 033837

[15] Barnett S M and Allen L 1994 Orbital angular momentum and nonparaxial light beams *Opt. Commun.* **110** 670–8

[16] van Enk S J and Nienhuis G 1994 Spin and orbital angular momentum of photons *Europhys. Lett.* **25** 497–501

[17] Santamato E 2004 Photon orbital angular momentum: problems and perspectives *Fortschr. Phys.* **52** 1141–53
[18] Bliokh K Y, Alonso M A, Ostrovskaya E A and Aiello A 2010 Angular momenta and spin–orbit interaction of nonparaxial light in free space *Phys. Rev. A* **82** 063825
[19] Coles M M and Andrews D L 2013 Photonic measures of helicity: optical vortices and circularly polarized reflection *Opt. Lett.* **38** 869–71
[20] Andrews D L, Dávila Romero L C and Babiker M 2004 On optical vortex interactions with chiral matter *Opt. Commun.* **237** 133–9
[21] Allen L, Beijersbergen M W, Spreeuw R J C and Woerdman J P 1992 Orbital angular momentum of light and the transformation of Laguerre–Gaussian laser modes *Phys. Rev. A* **45** 8185–9
[22] Nye J F and Berry M V 1974 Dislocations in wave trains *Proc. R. Soc. A* **336** 165–90
[23] Coullet P, Gil L and Rocca F 1989 Optical vortices *Opt. Commun.* **73** 403–8
[24] Dávila Romero L C, Andrews D L and Babiker M 2002 A quantum electrodynamics framework for the nonlinear optics of twisted beams *J. Opt. B: Quantum Semiclass. Opt.* **4** S66–72
[25] Bouchal Z and Olivík M 1995 Non-diffractive vector Bessel beams *J. Mod. Opt.* **42** 1555–66
[26] McGloin D and Dholakia K 2005 Bessel beams: diffraction in a new light *Contemp. Phys.* **46** 15–28
[27] Mazilu M, Stevenson D J, Gunn-Moore F and Dholakia K 2010 Light beats the spread: 'non-diffracting' beams *Laser Photon. Rev.* **4** 529–47
[28] Aiello A and Agarwal G S 2014 Wave-optics description of self-healing mechanism in Bessel beams *Opt. Lett.* **39** 6819–22
[29] Lóxpez-Mariscal C, Gutiérrez-Vega J C, Milne G and Dholakia K 2006 Orbital angular momentum transfer in helical Mathieu beams *Opt. Express* **14** 4182–7
[30] Alpmann C, Bowman R, Woerdemann M, Padgett M and Denz C 2010 Mathieu beams as versatile light moulds for 3D micro particle assemblies *Opt. Express* **18** 26084–91
[31] Bradshaw D S and Andrews D L 2017 Manipulating particles with light: radiation and gradient forces *Eur. J. Phys.* **38** 034008
[32] Andrews D L and Bradshaw D S 2018 *Introduction to Photon Science and Technology* (Bellingham, WA: SPIE Press)
[33] Andrews D L and Allcock P 2002 *Optical Harmonics in Molecular Systems* (Weinheim: Wiley-VCH) p. 10
[34] Babiker M, Power W L and Allen L 1994 Light-induced torque on moving atoms *Phys. Rev. Lett.* **73** 1239–42
[35] He H, Friese M E J, Heckenberg N R and Rubinsztein-Dunlop H 1995 Direct observation of transfer of angular-momentum to absorptive particles from a laser-beam with a phase singularity *Phys. Rev. Lett.* **75** 826–9
[36] Friese M E J, Enger J, Rubinsztein-Dunlop H and Heckenberg N R 1996 Optical angular-momentum transfer to trapped absorbing particles *Phys. Rev. A* **54** 1593–6
[37] Simpson N B, Allen L and Padgett M J 1996 Optical tweezers and optical spanners with Laguerre–Gaussian modes *J. Mod. Opt.* **43** 2485–91
[38] Simpson N B, Dholakia K, Allen L and Padgett M J 1997 Mechanical equivalence of spin and orbital angular momentum of light: an optical spanner *Opt. Lett.* **22** 52–4
[39] Roichman Y, Grier D G and Zaslavsky G 2007 Anomalous collective dynamics in optically driven colloidal rings *Phys. Rev. E* **75** 020401

[40] Molloy J E and Padgett M J 2002 Lights, action: optical tweezers *Contemp. Phys.* **43** 241–58
[41] Padgett M J and Bowman R 2011 Tweezers with a twist *Nat. Photonics* **5** 343–8
[42] Padgett M J 2014 Light's twist *Proc. R. Soc. A* **470** 0633
[43] Otte E and Denz C 2020 Optical trapping gets structure: structured light for advanced optical manipulation *Appl. Phys. Rev.* **7** 041308
[44] Yang Y, Ren Y, Chen M, Arita Y and Rosales-Guzmán C 2021 Optical trapping with structured light: a review *Adv. Photon.* **3** 034001
[45] Paterson L, MacDonald M P, Arlt J, Sibbett W, Bryant P E and Dholakia K 2001 Controlled rotation of optically trapped microscopic particles *Science* **292** 912–4
[46] Donato M G, Brzobohatý O, Simpson S H, Irrera A, Leonardi A A, Lo Faro M J, Svak V, Maragò O M and Zemánek P 2019 Optical trapping, optical binding, and rotational dynamics of silicon nanowires in counter-propagating beams *Nano Lett.* **19** 342–52

IOP Publishing

Optical Nanomanipulation (Second Edition)

David L Andrews and David S Bradshaw

Chapter 10

Structured light: particle steering and traction

Initially, optical manipulation with structured light mainly centred upon Laguerre–Gaussian beams, for example where ultracold atoms can be guided along the beam axis. However, new schemes continue to arise such as particle steering with other forms of wavefront-structured light such as Airy beams, and optical traction effects achieved using solenoid or Bessel beams. All of these systems have applications in microparticle manipulation, although it is widely expected that downscaling to nanoscale sizes is achievable. Such techniques rely on control of the detailed structuring of the trapping beam to produce the novel effects.

10.1 Particle steering

In the realm of atomic traps, the zero-intensity singularity at the core of Laguerre–Gaussian and Bessel beams provides an opportunity to channel atoms along the beam axis. This application takes advantage of the discrete character of electronic levels in atoms, such that their polarisability has a different sign on either side of a resonant wavelength, as we observed in chapter 8—see equation (8.1). Blue detuning can thus confine atoms, once introduced into the axis of the beam, to travel in a manner constrained by the higher intensities that surround them on either side [1, 2].

Another intriguing form of particle steering involves Airy beams: these are accelerating, non-diffracting beams whose paths of highest intensity do not propagate in a straight line [3–6]. Such beams have trajectories that can be engineered, for example, in a parabolic curve in the simplest form. Analogous to Bessel beams, this type of laser light contains self-healing properties as conveyed by figure 10.1. A particle within an Airy beam is shown to undergo ballistic dynamics, similar to projectiles acting under gravity [7]. Based on this concept a 'snowblower' effect has been realised [8], a description of which now follows.

In the 'snowblower' effect, so-termed by analogy to the propulsion of snow particles upwards and sideways (in an arc due to gravity), a sample of particles are exposed to an Airy beam. In such a system, the sample is typically separated into

doi:10.1088/978-0-7503-4191-2ch10

Figure 10.1. Self-healing properties of the Airy beam. (a) Ray tracing picture where black lines represent rays that are blocked by the obstacle (green horizontal line) and the red lines are rays that continue to propagate and reconstruct the Airy beam after some distance. (b) Electromagnetic field representation, in which the greyscale depicts the intensity profile of the reconstructing Airy beam, after an obstacle (pink line) is positioned in front of the first lobe. Adapted with permission of Wiley, from reference [3], copyright 2010.

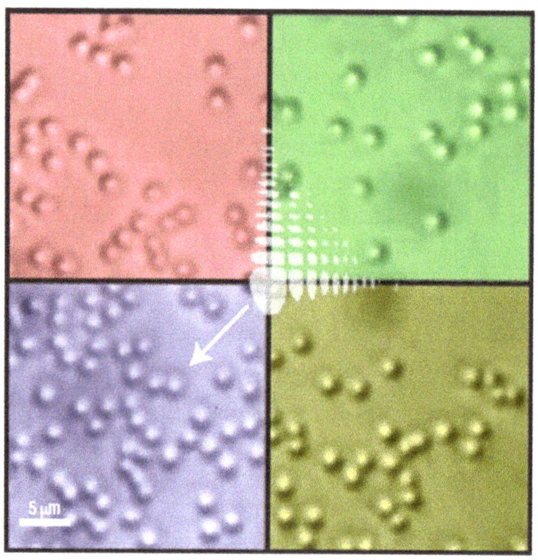

Figure 10.2. Demonstration of the 'snowblower' effect: a sample of colloids transfer from one compartment to another following exposure to an Airy beam for two minutes. The transverse field distribution of the Airy beam (white pattern) and the direction of particle propagation (white arrow) are overlaid; the micrograph is divided into four (coloured) sections. Reprinted by permission from Springer Nature: Nature. *Nature Photonics*, from reference [8], copyright 2008.

four compartments and the peak intensity of the asymmetric Airy beam is centrally positioned. The transverse field distribution of such a beam comprises a high intensity, central peak and a number of nearby lobes (with intensities that decay with increasing distance from the centre, as depicted by figure 10.2). The laser light attracts the particles to the high intensity portion of the beam via the gradient force and, subsequently, propels them along a parabolic path due to the radiation force of

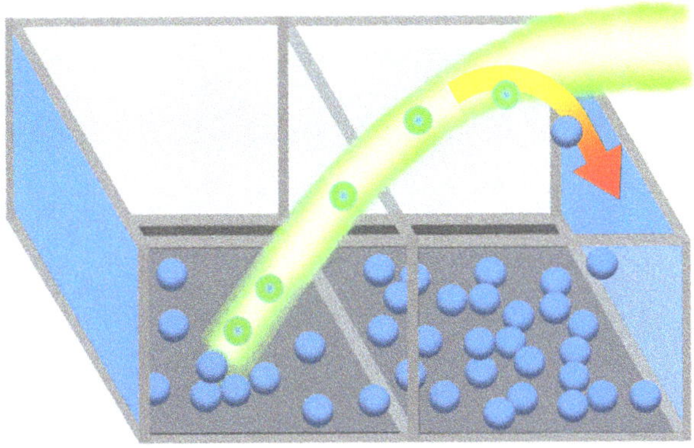

Figure 10.3. Artist's impression of the 'snowblowing' effect. Particles are transferred between compartments because the Airy beam (in green) acts as an optical guide.

the curved beam. The steering continues, typically for a distance of around 75 μm, until the beam becomes 'smeared out' and the particles are released. The result, as shown in figures 10.2 and 10.3, is that the Airy beam launches the particles from one compartment to another.

10.2 Tractor beams

As we established in chapter 5, the radiation force of a trapping beam generally pushes the particles in the same direction as the propagating light. However, the possibility that light can attract objects towards its source, in the opposite direction to every intuition, has interested researchers for many decades [9]. This phenomenon has been branded a *pulling force* or a *tractor beam*, a concept that originates from the science fiction genre. Over the last 10–15 years, by tailoring the properties of certain laser beam configurations, the prospect of this intriguing ability has evolved into a viable emerging field [10, 11]. The first experimental set-up involved two interfering Gaussian laser beams with wavevectors $\mathbf{k}_1$ and $\mathbf{k}_2$; this configuration is shown in figure 10.4(a). The pulling force arises for specific combinations of parameters, where most of the incident photons are scattered in the forward direction. Based on the principles of action and reaction, it is possible for a transfer of light–matter momentum to produce a backward movement of a particle [12]. This effect is highly sensitive to the polarisation of the incident beams, to the point that switching from a pulling to a pushing force is attainable by altering the beam polarisations—as depicted in figure 10.4(b) and (c).

A structured form of laser light that may act as an optical tractor is known as a *solenoid beam*, which is in a sense an optical analogue to the Archimedes' screw. Such structured light is characterised by an intensity maximum that spirals round the beam axis—as shown in figure 10.5(a). The upstream movement of a trapped particle, depicted in figure 10.5(b), occurs due to a combination of the gradient and the radiation force of the solenoid beam, i.e. a component of the combined force is directed counter to the light propagation [13, 14]. (A paraxial Gaussian beam cannot

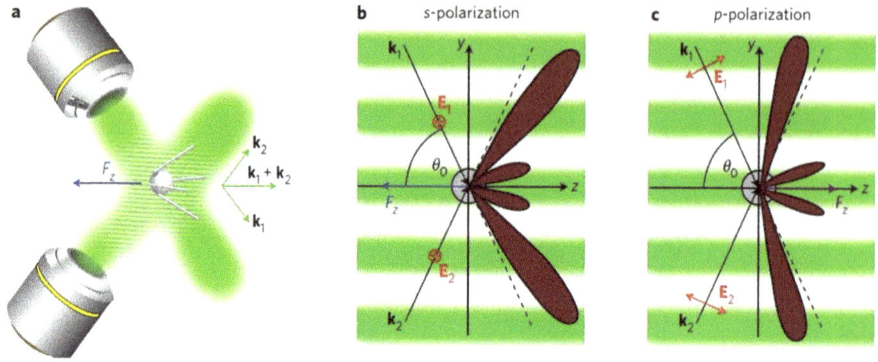

Figure 10.4. (a) Geometry of two Gaussian beams propagating with wave vectors $\mathbf{k}_1$ and $\mathbf{k}_2$. The pulling force, within a light fringe (along the z-axis), is opposite to the direction of the combined wave vectors, $\mathbf{k}_1 + \mathbf{k}_2$. (b) Illustration of the light scattering of s-polarized beams with electric fields, $\mathbf{E}_1$ and $\mathbf{E}_2$, oriented along the x-axis. Most of the light is forward-scattered, producing an optical pulling force. (c) Geometry of p-polarized beams, which generates a conventional pushing force. Reprinted by permission from Springer Nature: Nature. *Nature Photonics*, from reference [12], copyright 2013.

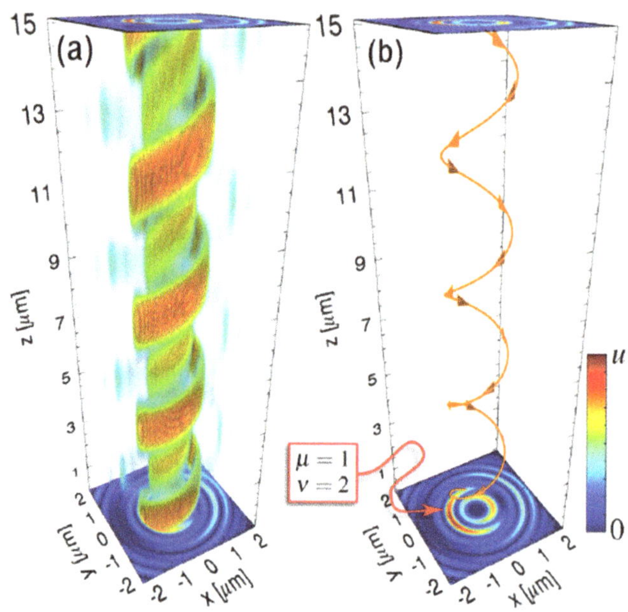

Figure 10.5. (a) Intensity distribution of a right-handed solenoidal tractor beam, with a wavelength of 532 nm, propagating along the $+z$-direction through water. (b) Trajectory of a 50 nm diameter silica sphere moving along the tractor beam in the $-z$-direction. The colour bar indicates the local intensity, u. Reprinted with permission from reference [13], copyright 2016, by the American Physical Society.

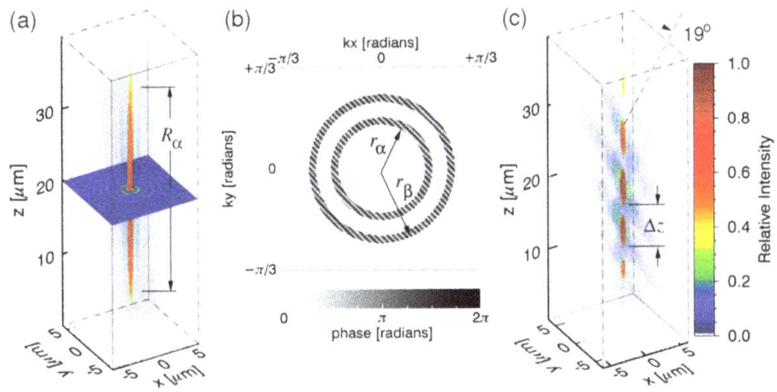

Figure 10.6. (a) Volumetric reconstruction of a Bessel beam projected with a ring-like hologram. (b) Phase hologram encoding an optical conveyor. Diagonal blazing tilts the projected conveyor away from the optical axis. (c) Volumetric reconstruction of the beam projected by the hologram in (b). Reprinted with permission from reference [17], copyright 2012, by the American Physical Society.

serve in such capacity since all its momentum is directed along the axis of propagation.) Another form of tractor beam is known as an *optical conveyor*, which can selectively transport illuminated objects either upstream or downstream along the beam axes; such systems can be generated by a superposition of coaxial Bessel beams [15–18]. These conveyors have periodic intensity patterns along their axis (figure 10.6), as a result the gradient force may override the push of the radiation force. Varying the relative phase of the beams will shift the axial positions of the optical traps. In this way, it is possible to control the passage of the trapped objects downstream or upstream.

10.3 Surface plasmon optical vortex

We now turn to an interesting development of the optical trapping technique involving counter-propagating Laguerre–Gaussian beams, which was introduced in the previous chapter. Here, in contrast, attention focuses on the motion of atoms in a set-up that engenders the formation of surface plasmon optical vortices.

The burgeoning field of plasmonics was briefly mentioned in chapter 5, a fuller explanation is now provided. Plasmonics is one of the most active areas of current research and development in the science of optics and optical materials. Most of these studies concern the light-induced excitation of *surface plasmons*: quantised charge oscillations that can form on the surface of a metal or at a metal interface with a dielectric solid. The strongly enhanced electric fields that such plasmons produce at boundaries and edges leads into a rich variety of applications [19–27]. On smooth metals, surface plasmons share quantum propagation features of both electrons and photons, and their character is strongly influenced by the specific electronic properties of the metal at optical frequencies. Indeed, this is reflected in

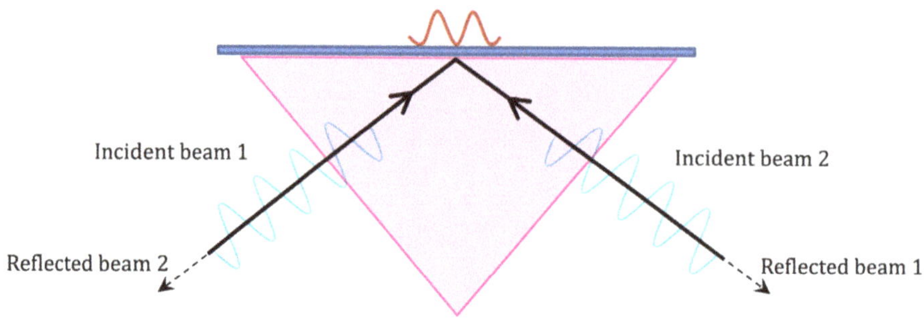

Figure 10.7. Plasmonic scheme based on the Kretschmann prism (pink triangle) configuration. Total internal reflection of two counter-propagating Laguerre–Gaussian beams (black arrows), at a planar dielectric interface with a metallic film (blue rectangle), creates a surface plasmon optical vortex (red wavy line).

another common designation, 'surface plasmon polaritons'—see, for example, chapter 12 of reference [28]. Under normal circumstances, the surface plasmon of a smooth metal cannot be excited directly with light from a vacuum, because of a sizeable mismatch between the momentum conveyed by a free-space photon and the momentum of a plasmon with the corresponding frequency. However, this problem can be overcome by arranging for light to impinge upon a metallic film by internal reflection inside a prism, wherein the photon momentum is larger than its vacuum value.

It transpires that localised non-propagating surface plasmons can be formed by exploiting the properties of counter-propagating optical vortex light, since such beams can instigate plasmonic excitation [29]. A suitable set-up is shown in figure 10.7, based on a Kretschmann prism configuration, where counter-propagating Laguerre–Gaussian beams undergo total internal reflection at a common position on a prism face—this results in the formation of a surface plasmon optical vortex [30–34]. Whereas surface plasmons generally have a propagating character associated with linear momentum along the surface, the two-beam configuration provides for a cancellation of equal and opposite surface components so that the excitation is sustained at a fixed (localised) position, where the input beams meet. Conversely, however, when those beams have orbital angular momenta of equal and opposite sign, their positive addition provides a combined electrodynamic torque that drives the plasmon excitation, resulting in vortex behaviour. The transverse field distribution and angular momentum of the resulting plasmon can be visualised with near-field scanning optical microscopy [35]. Following formation of the surface plasmon optical vortex [36], any atom within the vicinity of the metallic surface may be trapped in the high intensity part of the optical vortex, due to the gradient force, and will undergo motion in a plane parallel to the surface as a result of the radiation force—a typical atom trajectory is shown in figure 10.8(a),(b). It is important to note

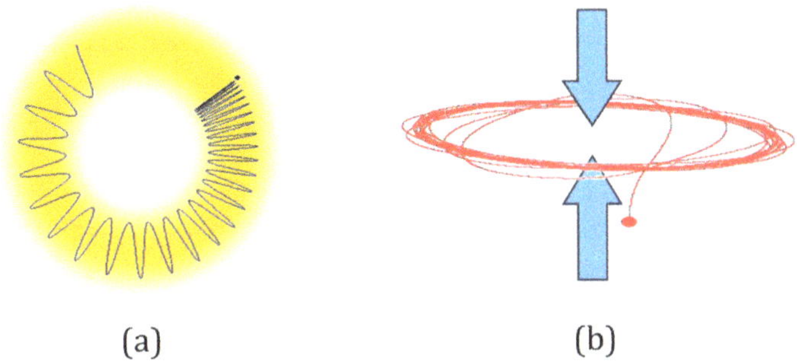

Figure 10.8. (a) Typical trajectory (grey line) of an atom within a surface plasmon optical vortex. The scheme is based on two red-detuned, counter-propagating Laguerre–Gaussian beams: one with $l = 1$ and the other $l = -1$, and $p = 0$. The dot represents the initial position of the atom, which then accelerates around the beam, zigzagging in and out as it is subjected to the radial potential well. (b) At higher intensities, the steepness of the radial potential well results in more rapid circulation of the atom in a well-confined ring (red line).

that any theory relating to this type of trapping belongs to the strong coupling regime, where light–matter interactions are large. Therefore, the perturbation methods introduced in chapter 2 cannot be applied to this plasmonic system.

References

[1] Arlt J, Hitomi T and Dholakia K 2000 Atom guiding along Laguerre–Gaussian and Bessel light beams *Appl. Phys. B* **71** 549–54

[2] Rhodes D P, Gherardi D M, Livesey J, McGloin D, Melville H, Freegarde T and Dholakia K 2006 Atom guiding along high order Laguerre–Gaussian light beams formed by spatial light modulation *J. Mod. Opt.* **53** 547–56

[3] Mazilu M, Stevenson D J, Gunn-Moore F and Dholakia K 2010 Light beats the spread: 'non-diffracting' beams *Laser Photon. Rev.* **4** 529–47

[4] Siviloglou G A, Broky J, Dogariu A and Christodoulides D N 2007 Observation of accelerating Airy beams *Phys. Rev. Lett.* **99** 213901

[5] Broky J, Siviloglou G A, Dogariu A and Christodoulides D N 2008 Self-healing properties of optical Airy beams *Opt. Express* **16** 12880–91

[6] Dholakia K 2008 Optics: against the spread of the light *Nature* **451** 413

[7] Siviloglou G A, Broky J, Dogariu A and Christodoulides D N 2008 Ballistic dynamics of Airy beams *Opt. Lett.* **33** 207–9

[8] Baumgartl J, Mazilu M and Dholakia K 2008 Optically mediated particle clearing using Airy wavepackets *Nat. Photonics* **2** 675–8

[9] Dogariu A, Sukhov S and Sáenz J 2013 Optically induced 'negative forces' *Nat. Photonics* **7** 24–7

[10] Ding W, Zhu T, Zhou L-M and Qiu C-W 2019 Photonic tractor beams: a review *Adv. Photon.* **1** 024001

[11] Li H, Cao Y, Zhou L-M, Xu X, Zhu T, Shi Y, Qiu C-W and Ding W 2020 Optical pulling forces and their applications *Adv. Opt. Photon.* **12** 288–366

[12] Brzobohatý O, Karásek V, Šiler M, Chvátal L, Čižmár T and Zemánek P 2013 Experimental demonstration of optical transport, sorting and self-arrangement using a 'tractor beam' *Nat. Photonics* **7** 123–7
[13] Yevick A, Ruffner D B and Grier D G 2016 Tractor beams in the Rayleigh limit *Phys. Rev. A* **93** 043807
[14] Lee S-H, Roichman Y and Grier D G 2010 Optical solenoid beams *Opt. Express* **18** 6988–93
[15] Čižmár T, Garcés-Chávez V, Dholakia K and Zemánek P 2005 Optical conveyor belt for delivery of submicron objects *Appl. Phys. Lett.* **86** 174101
[16] Čižmár T, Kollárová V, Bouchal Z and Zemánek P 2006 Sub-micron particle organization by self-imaging of non-diffracting beams *New J. Phys.* **8** 43
[17] Ruffner D B and Grier D G 2012 Optical conveyors: a class of active tractor beams *Phys. Rev. Lett.* **109** 163903
[18] Ruffner D B and Grier D G 2014 Universal, strong and long-ranged trapping by optical conveyors *Opt. Express* **22** 26834–43
[19] Zayats A V and Smolyaninov I I 2003 Near-field photonics: surface plasmon polaritons and localized surface plasmons *J. Opt. A: Pure Appl. Opt.* **5** S16–50
[20] Ozbay E 2006 Plasmonics: merging photonics and electronics at nanoscale dimensions *Science* **311** 189–93
[21] Dragoman M and Dragoman D 2008 Plasmonics: applications to nanoscale terahertz and optical devices *Prog. Quant. Electron.* **32** 1–41
[22] Kawata S, Inouye Y and Verma P 2009 Plasmonics for near-field nano-imaging and superlensing *Nat. Photonics* **3** 388–94
[23] Atwater H A and Polman A 2010 Plasmonics for improved photovoltaic devices *Nat. Mater.* **9** 205–13
[24] Schuller J A, Barnard E S, Cai W, Jun Y C, White J S and Brongersma M L 2010 Plasmonics for extreme light concentration and manipulation *Nat. Mater.* **9** 193–204
[25] Kim J 2012 Joining plasmonics with microfluidics: from convenience to inevitability *Lab Chip* **12** 3611–23
[26] Zheng Y B, Kiraly B, Weiss P S and Huang T J 2012 Molecular plasmonics for biology and nanomedicine *Nanomedicine* **7** 751–70
[27] Sonntag M D, Klingsporn J M, Zrimsek A B, Sharma B, Ruvuna L K and Van Duyne R P 2014 Molecular plasmonics for nanoscale spectroscopy *Chem. Soc. Rev.* **43** 1230–47
[28] Novotny L and Hecht B 2006 *Principles of Nano-Optics* (Cambridge: Cambridge University Press)
[29] Tan P S, Yuan X-C, Lin J, Wang Q, Mei T, Burge R E and Mu G G 2008 Surface plasmon polaritons generated by optical vortex beams *Appl. Phys. Lett.* **92** 111108
[30] Lembessis V E, Babiker M and Andrews D L 2009 Surface optical vortices *Phys. Rev. A* **79** 011806
[31] Andrews D L, Babiker M, Lembessis V E and Al-Awfi S 2010 Surface plasmons with phase singularities and their effects on matter *Phys. Status Solidi Rapid Res. Lett.* **4** 241–3
[32] Lembessis V E, Al-Awfi S, Babiker M and Andrews D L 2011 Surface plasmon optical vortices and their influence on atoms *J. Opt.* **13** 064002
[33] Al-Awfi S 2013 Formation of a plasmonic surface optical vortex by evanescent bessel light *Plasmonics* **8** 529–36

[34] Babiker M, Andrews D L and Lembessis V E 2018 Atoms in complex twisted light *J. Opt.* **21** 013001
[35] Shen Z, Hu Z J, Yuan G H, Min C J, Fang H and Yuan X C 2012 Visualizing orbital angular momentum of plasmonic vortices *Opt. Lett.* **37** 4627–9
[36] Spektor G *et al* 2017 Revealing the subfemtosecond dynamics of orbital angular momentum in nanoplasmonic vortices *Science* **355** 1187–91

IOP Publishing

Optical Nanomanipulation (Second Edition)

David L Andrews and David S Bradshaw

Chapter 11

Optofluidics: lab-on-a-chip mixing and actuating flow

The optical manipulation and control of microfluids, or particles within them, is increasingly important for effective operations in lab-on-a-chip devices. For example, an optical stirring technique based on a Laguerre–Gaussian beam can solve the problem of insufficient turbulent flow in micro-flow systems, thereby enabling any otherwise stifled chemical reactions. In addition, optically trapped particles can act as microfluidic pumps or valves to open and close certain micro-channels in the device. The ability to control fluid flows at reduced length and time scales has now become pressing. This will require novel nanophotonic approaches, the initial steps towards creating them has now begun.

11.1 Overview

As we have seen, the optical forces that light exerts on matter have a variety of origins, yet in terms of application they fall naturally into essentially two groups. There are those in which a relatively insignificant resistance to optically applied forces leads to motion that is constrained only, or primarily, by a positional (or angular) variation in the associated force field; optical tweezers is an obvious example. Where such local variation is less pronounced—or is at least monotonic—then sustained motion may ensue, as for example occurs in optical methods of particle separation and cell sorting. All such cases, where the responsible forces are determined only by spatial position, satisfy one of the several alternative criteria for a *dynamically conservative system*.

However, there are also systems, especially fluids, in which the aim of achieving optically driven motion is frustrated by other local forces: friction or viscosity are the most typical manifestation. Thus, particles in liquid suspension driven ‘downstream’ along the axis of a laser beam by radiation pressure attain a terminal velocity,

doi:10.1088/978-0-7503-4191-2ch11

typically of the order of 10 μm s^{-1}, as a result of velocity-dependent resistance. Some of the energy content of the driving beam can, therefore, be spent in sustaining constant velocity motion through the host medium. Thermal effects that produce rapidly fluctuating, stochastic variations in local force fields can also modify and constrain uniform motion. Nonetheless, for the system as a whole—that is, the particles of interest, the light and the surrounding medium—energy is, of course, always conserved.

11.2 Optical manipulation in microfluidics

Using light to manipulate fluids, or particles within fluid media, is a concept described by the umbrella term *optofluidics*. This idea is especially important for lab-on-a-chip devices, in which laboratory functions can be implemented upon a chip with centimetre- or millimetre-scale dimensions (figure 11.1). Instruments of this type are useful for diagnostics, sensing, testing, pathology and drug discovery. Optical tweezing has an important role in microfluidic systems, such as the mixing of minuscule amounts of chemicals, the actuating of flow and the sorting of particles. (The latter is covered in chapter 7 for cellular bodies; however, the sorting of colloids or similar particles is equally possible in such systems.) It is noteworthy that the practicality of nanofluidics, beyond single molecule sensing, encounters formidable difficulties due to the exceptionally slow volumetric flow of any liquid confined to nanoscale channels [1].

Liquid flow on the microscale generally exhibits laminar (non-turbulent) flow. This arises due to the high resistance against the flow of a typical liquid confined to the microscale; such fluids have a low *Reynolds number*. (The latter is defined as the ratio of inertial forces to viscous forces.) In application to lab-on-a-chip devices,

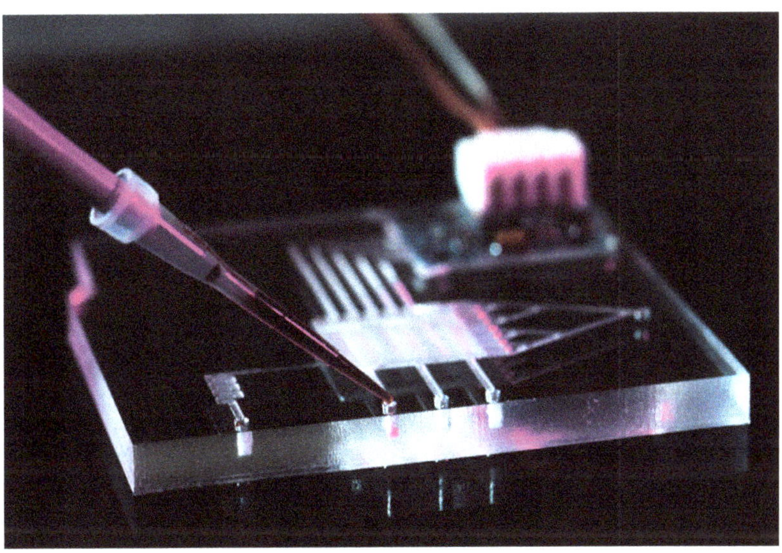

Figure 11.1. Image of a typical lab-on-a-chip device. Image from iStock.com/Luchschen.

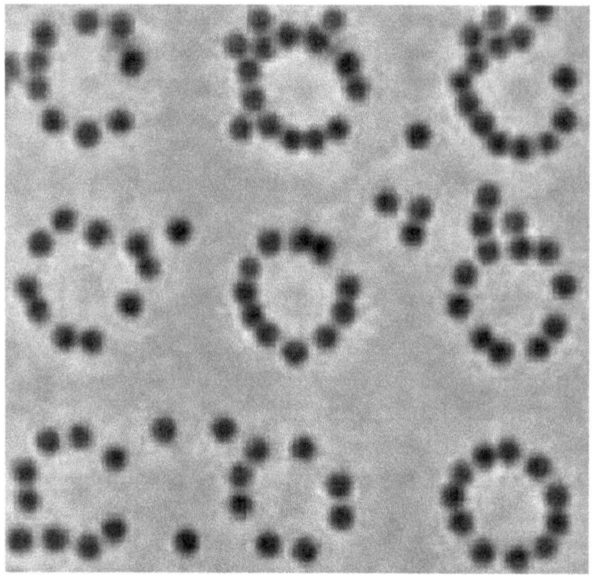

Figure 11.2. Colloidal polystyrene spheres (800 nm in diameter) trapped and circulating within a 3×3 array of Laguerre–Gaussian beams, each with $l = 15$. Reprinted from reference [2], copyright 2002, with permission from Elsevier.

laminar flow is ideal for transferring minuscule quantities of reagents to a reaction site and the sorting of particles. However, the lack of turbulent flow means that the mixing of the reactants is difficult [3–5]. This problem can be overcome by a phenomenon introduced in chapter 9: optically trapped (chemically inert) particles that circulate round a ring of a Laguerre–Gaussian beam could induce the turbulence required for a chemical reaction. Moreover, amplification of such stirring may be achieved by using a holographic optical tweezers set-up that projects multiple optical vortices [6–8] (for example, the 3×3 array given by figure 11.2)—interestingly, using the same set of principles, the circulating particles may act as an optomechanical pump to create a flow [9]. Outside of this holographic system, where particles are rotated via OAM transfer, other mixing and pumping schemes based on optical tweezing have been proposed. Here, momentum is exchanged from a Gaussian beam to a microgear [10, 11], a micropropeller [12] or a calcite crystal [13].

Instead of rotating the microparticles, another optical pump technique involves the translation of particles back and forth in a cooperative manner, as shown by figure 11.3(a). Fabrication of these pumps requires an independent optical trap for each of the microparticles, so that sine wave propagation can be produced to drive the flow of the microscale liquid [14]. An optical valve to actively direct particles into one of two exit channels, as depicted in figure 11.3(b), has also been demonstrated. In this case, the particles that constitute the valve structure are rotated about a swivel point by an optical trap, so that either the upper or lower channel is sealed

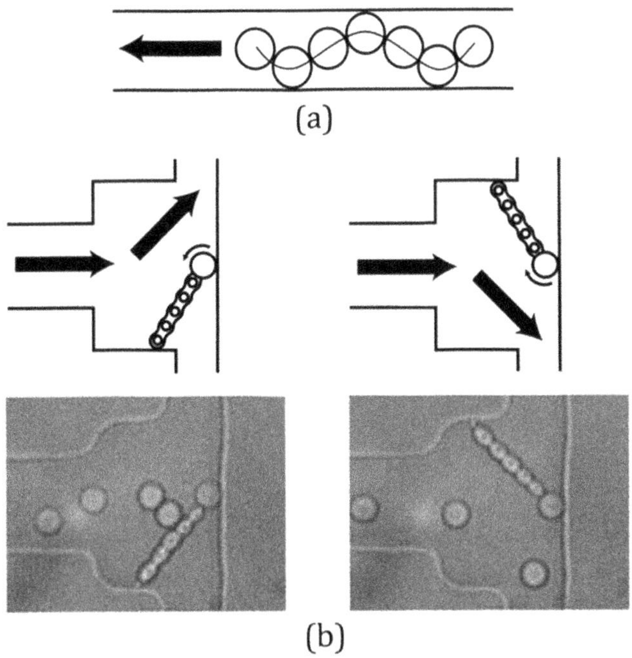

Figure 11.3. (a) Optical pump drives fluid motion through a sinusoidal rippling of a colloidal chain; (b) a valve constructed of colloidal particles in microfluidic channels. Both are activated with optical tweezers, and the direction of fluid flow is indicated by the arrows. The colloidal valve flap is flipped with an optical tweezer and directs particles either upward (left) or downward (right). Actual image of the colloidal peristaltic pump, corresponding to diagram (a), appears in reference [14] and image (b) is reprinted from reference [15], copyright 2002, with the permission of AIP Publishing.

and the flow is directed towards the open channel [15]. Another useful technique, based on a microfluidic system that uses optical tweezers, is known as a beam steerer. This works on the principle that an optically trapped microfluidic element, e.g. a silica microsphere, acts as a lens and refracts by varying degrees a signal beam as the particles moves across it [16, 17].

In recent years, in the continually expanding field of microfluidics, the ability to control fluid flows at reduced length and time scales (and, thus, faster reaction rates) has become increasingly important. It is anticipated that this facility will require novel nanophotonic approaches, in the form of 'nanomixers', to expedite and significantly enhance the capabilities of existing microfluidic techniques. The initial steps to achieve this aim have already begun: simulations have been performed on a scheme that mixes fluids within microfluidic channels via a silicon nanocube antenna, suspended gold nanoparticles and incident circularly polarised light (figure 11.4) [18]. In short, while lab-on-a-chip devices remain in their infancy, it is expected that the refining of techniques for future designs will be heavily reliant on optical nanomanipulation effects.

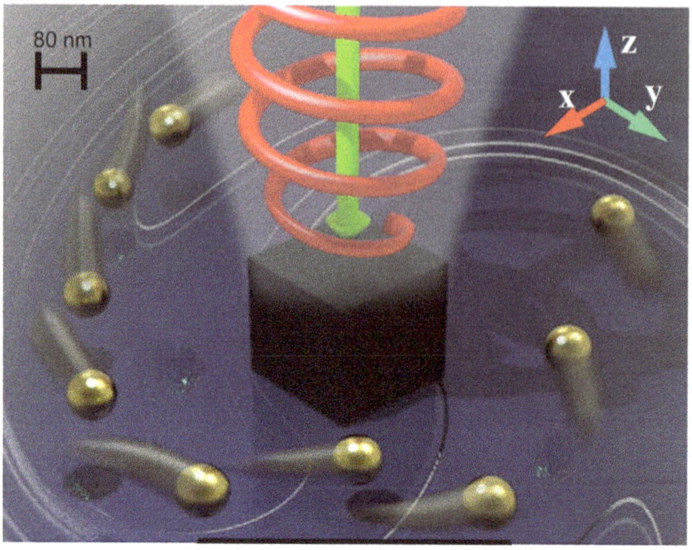

Figure 11.4. Artist's impression of an optical nanomixing scheme. A 200 nm silicon nanocube antenna (black cube) immersed in water is illuminated by a circularly polarized laser beam (red spiral). The scattered field carries a non-zero tangential component, inducing gold nanoparticles (80 nm in diameter) to spiral around the nanocube. Fluid mixing arises due to the viscous forces generated by the motion of the gold nanoparticles. Reprinted with permission of Wiley, from reference [18], copyright 2020.

References

[1] Erickson D, Serey X, Chen Y-F and Mandal S 2011 Nanomanipulation using near field photonics *Lab Chip* **11** 995–1009

[2] Curtis J E, Koss B A and Grier D G 2002 Dynamic holographic optical tweezers *Opt. Commun.* **207** 169–75

[3] Lee C-Y, Chang C-L, Wang Y-N and Fu L-M 2011 Microfluidic mixing: a review *Int. J. Mol. Sci.* **12** 3263–87

[4] Ward K and Fan Z H 2015 Mixing in microfluidic devices and enhancement methods *J. Micromech. Microeng.* **25** 094001

[5] Lee C-Y, Wang W-T, Liu C-C and Fu L-M 2016 Passive mixers in microfluidic systems: a review *Chem. Eng. J.* **288** 146–60

[6] Grier D G 2003 A revolution in optical manipulation *Nature* **424** 810–6

[7] Padgett M and Di Leonardo R 2011 Holographic optical tweezers and their relevance to lab on chip devices *Lab Chip* **11** 1196–205

[8] Mohanty S 2012 Optically-actuated translational and rotational motion at the microscale for microfluidic manipulation and characterization *Lab Chip* **12** 3624–36

[9] Ladavac K and Grier D G 2004 Microoptomechanical pumps assembled and driven by holographic optical vortex arrays *Opt. Express* **12** 1144–9

[10] Neale S L, MacDonald M P, Dholakia K and Krauss T F 2005 All-optical control of microfluidic components using form birefringence *Nat. Mater.* **4** 530–3

[11] Metzger N K, Mazilu M, Kelemen L, Ormos P and Dholakia K 2011 Observation and simulation of an optically driven micromotor *J. Opt.* **13** 044018

[12] Galajda P and Ormos P 2001 Complex micromachines produced and driven by light *Appl. Phys. Lett.* **78** 249–51
[13] Friese M E J, Nieminen T A, Heckenberg N R and Rubinsztein-Dunlop H 1998 Optical alignment and spinning of laser-trapped microscopic particles *Nature* **394** 348–50
[14] Terray A, Oakey J and Marr D W M 2002 Microfluidic control using colloidal devices *Science* **296** 1841–4
[15] Terray A, Oakey J and Marr D W M 2002 Fabrication of linear colloidal structures for microfluidic applications *Appl. Phys. Lett.* **81** 1555–7
[16] Domachuk P, Cronin-Golomb M, Eggleton B J, Mutzenich S, Rosengarten G and Mitchell A 2005 Application of optical trapping to beam manipulation in optofluidics *Opt. Express* **13** 7265–75
[17] Domachuk P, Omenetto F G, Eggleton B J and Cronin-Golomb M 2007 Optofluidic sensing and actuation with optical tweezers *J. Opt. A: Pure Appl. Opt.* **9** S129–33
[18] Canós Valero A, Kislov D, Gurvitz E A, Shamkhi H K, Pavlov A A, Redka D, Yankin S, Zemánek P and Shalin A S 2020 Nanovortex-driven all-dielectric optical diffusion boosting and sorting concept for lab-on-a-chip platforms *Adv. Sci.* **7** 1903049

David L Andrews and David S Bradshaw

Chapter 12

Optical binding

Optical binding is a term applied to the formation of stable, non-contact arrangements of particles held in an optical trap. The presence of those particles in the laser light introduces a richly structured potential energy field. This results in an interplay of attractive and repulsive inter-particle forces producing well-separated local energy minima, such that not only linear but also two- and three-dimensional arrays can be supported in stable configurations. Such effects can be observed over a surprisingly diverse range of particle types and sizes. Among much else, they provide a basis for novel forms of optical self-assembly and nanoparticle sorting.

12.1 The nature of optical binding

Thus far, we have focused attention on the response of individual particles to optical forces. Provided each particle does not significantly occlude the light, then other particles within the same focal region might be anticipated to respond in a similar manner. However, for any such particles in close proximity, it emerges that new *inter-particle* forces, also of electromagnetic origin, may then additionally come into play. These non-contact forces can lead to local potential energy minima that are able to support the creation of a special, intrinsic type of optical lattice, operating over distances where so-called 'dispersion forces' apply. Wide-ranging studies of this effect have led to a recognition of many interesting observations [1]. To understand how these additional light-induced forces arise, the electrodynamic origin of dispersion forces first needs to be recognised.

12.2 The dispersion force: a comparison

Dispersion forces—also known as London forces [2–4], and principally represented by the Casimir–Polder interaction—are the forces that exist between any electrically neutral and essentially non-polar particles, beyond the 'contact' region where their individual electronic wavefunctions would significantly overlap. Typically characterised by a monotonic inverse-sixth power dependence on separation, such

doi:10.1088/978-0-7503-4191-2ch12

inter-particle forces of attraction operate over a surprisingly large range of scale—their importance is not limited to atoms or molecules, but even extend to the Hamaker interactions [5] that hold together ceramic microparticles. Indeed, these dispersion forces are responsible for the sustained integrity of most condensed phase matter; in their absence, repulsion between the outermost regions of electron density in separate particles would lead to instant material vaporisation. As Casimir and Polder showed [6], the distance dependence does eventually tail off into an inverse-seventh power, whose experimental proof verified that all such forces are ultimately produced by vacuum fluctuations. Associated with the latter, the fleeting presence of 'virtual' photons between the particles are responsible for the dispersion forces [7].

Returning to our main topic—and in clear distinction to the dispersion forces that occur whether or not any light is present—there are, in fact, additional forces that arise between particles, specifically connected with the presence of light. In many experiments these optically induced inter-particle forces prove to be much larger than the dispersion forces, and although they both diminish with distance, the former can still be significant at separations that are similar in magnitude to the optical wavelength. The effect is now commonly known by the term 'optical binding', coined by the authors of some of the earliest experimental work [8]. Nonetheless, as we shall see, the form of this interaction does not always have the quality of attraction that 'binding' usually signifies.

12.3 Theory of optical binding

Classical description. Before describing the quantum mechanism, it is helpful to reason through the origin of optical binding forces in essentially classical terms. Consider a pair of particles, in reasonably close proximity, that are irradiated by light of sufficient intensity. Each particle (which intrinsically has the property of polarisability—meaning that its constituent charges move in response to the electromagnetic radiation) will exhibit an oscillating dipole, due to the presence of the laser light. These oscillations will, however, generally differ in phase due to the difference in positioning of the two particles with respect to the light waves. The oscillating dipoles create secondary fields which are experienced by the partner particle. The result is mutual interaction between the particles which, depending on the inter-particle disposition, may be either in-phase or out-of-phase—leading to either repulsive or attractive forces, respectively. So we can anticipate that 'optical binding' forces will fall off with distance, cycling between progressively diminishing positive and negative values as the separation increases. This scheme of rationalisation is illustrated in figure 12.1.

Quantum description. A rigorous theory of optical binding calls for the deployment of quantum electrodynamical methods, wherein all inter-particle forms of interaction are mediated by 'virtual' photon exchange. First formulated by Thirunamachandran for specific application to molecules [9], this theory provides a definitive mechanism and a quantitative basis for characterising the effect, as qualitatively described above. The essential mechanism is pairwise forward Rayleigh scattering [10] (analogous to the gradient force of single-particle optical trapping),

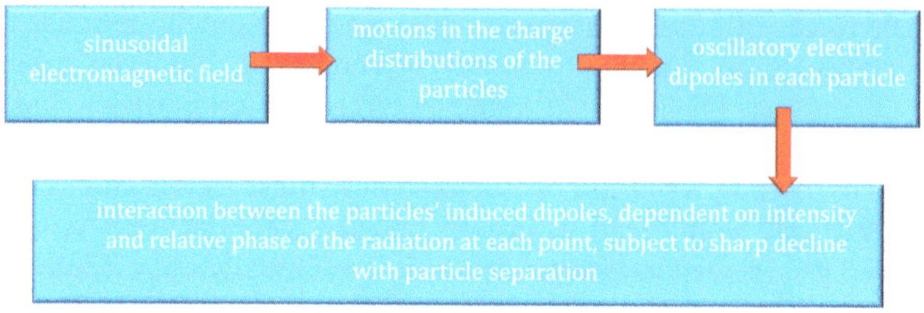

Figure 12.1. Outline of the classical description of optical binding.

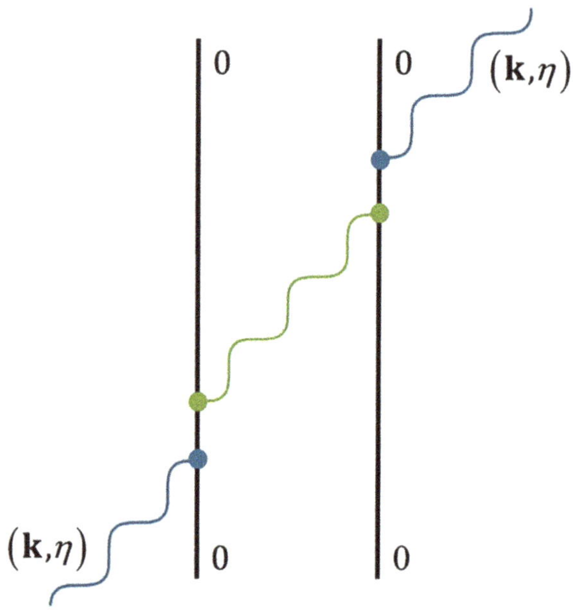

Figure 12.2. Feynman diagram illustrating optical binding. Blue and green wavy lines, respectively, represent 'real' and 'virtual' photons, and the dots denote light–matter interaction. The annihilated and created throughput beam ('real') photons are identical, and both particles begin and end in the ground state 0.

which involves the annihilation of one laser beam photon by one particle and re-creation of an identical photon, repopulating the beam, by the neighbour. The two events are causally linked by 'virtual' photon coupling, as shown by the Feynman diagram of figure 12.2. By summing the matrix elements corresponding to each such diagram, a pair interaction energy can be evaluated, whose relative position-dependence allows the evaluation of local forces. Remarkably, this theory proves to be applicable to much more than molecules, including dielectric particles of nanometre and micron size, and even extending to metal nanoparticles. (At the lower end of the scale, it also represents a mechanism for optical electrostriction—the bulk effect on atomic solids that we touched upon in chapter 3.) It emerges that this responsive force depends on a product of the polarisabilities for the interacting

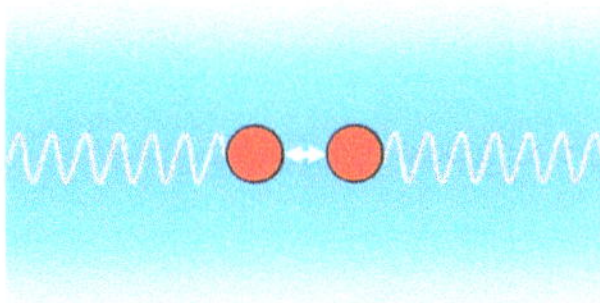

(a)

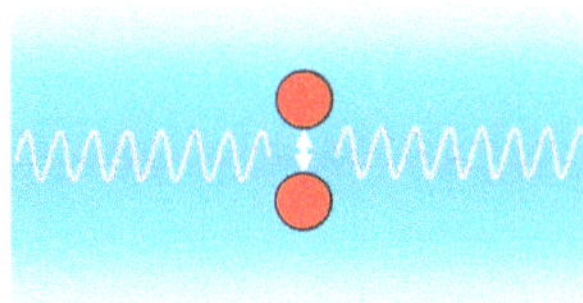

(b)

Figure 12.3. Diagrams depicting (a) longitudinal and (b) transverse optical binding. White double-headed arrows represent inter-particle interactions and the grey wavy lines denote the throughput of a Gaussian beam.

pair, and as such is much more amenable to experimental observation with relatively large and more polarisable components. This tallies with the quantum perspective, in which the net energies and forces for such composite particles result from a sum of pairwise interactions between the atomic-level constituents of each particle.

In addition to a rapidly declining dependence on distance, local forces also depend on the orientation of the pair displacement vector **s** with respect to the beam propagation vector **k** and polarisation vector **e**. Since optical binding involves forward Rayleigh scattering, where the initial and final system states are identical, the observable is a potential energy. Hence the key equations, for 'longitudinal' (where $\mathbf{s}\|\mathbf{k}$, $\mathbf{s}\perp\mathbf{e}$) and 'transverse' ($\mathbf{s}\perp\mathbf{k}$, $\mathbf{s}\|\mathbf{e}$) optical binding, respectively—as depicted in figure 12.3—are given as follows;

$$\Delta E = \left(\frac{I\alpha_A\alpha_B}{4\pi\varepsilon_0^2 cs^3}\right)\left(\left(1 - k^2s^2\right)\cos(2ks) + ks\sin(2ks)\right), \tag{12.1}$$

$$\Delta E = -\left(\frac{I\alpha_A\alpha_B}{2\pi\varepsilon_0^2 cs^3}\right)\left(\cos(ks) + ks\sin(ks)\right) \tag{12.2}$$

Here, α_A and α_B are respectively the scalar polarisabilities of particle A and B, and $s = |\mathbf{s}|$ is the separation distance between the two particles. The optical binding force is then determined by inserting equations (12.1) and (12.2) into $\mathbf{F} = -\partial(\Delta E)/\partial\mathbf{s}$; the intricate result (found elsewhere in a generalised form [10]) will show that optical binding has a short-range inverse-fourth power distance dependence, i.e. s^{-4}. In this chapter, representing most applications, the throughput beam is assumed Gaussian and monochromatic; the particles are assumed to be spherical. However, optical binding using structured light has also been examined [11–19], and the theory acquires additional complications when broadband radiation or anisotropic particles are entertained [20, 21].

How the potential energy varies with particle separation is illustrated in figure 12.4 for the two configurations; this clearly shows that the deeper potential wells, and hence stronger forces, are associated with particles positioned along the beam axis. The most striking aspect, in each case, is a series of separation distances at which the energy is a minimum, such that the optically induced force is neither attractive nor repulsive, but vanishes (since the gradient of the curve is zero). At each of these

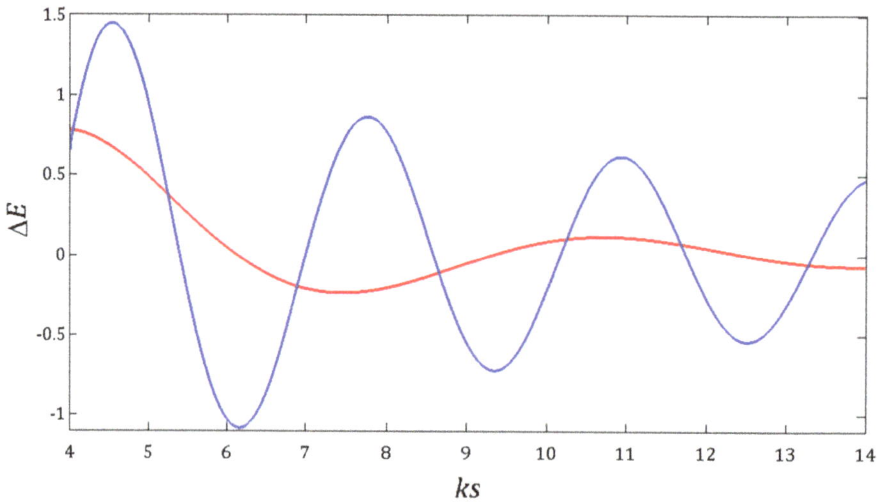

Figure 12.4. Graph of potential energy, ΔE, (with arbitrary scale) against pair separation measured in terms of ks, for longitudinal (blue line) and transverse (red line) optical binding; k is given a typical value of 3×10^{-7} m^{-1}. Notice that both oscillating curves are damped as the separation distance increases.

points, two particles can be held in a sustainable equilibrium position, well beyond material contact. In the case of the closest, deepest potential energy minimum the arrangement can be stable against small perturbations; indeed, by disturbing the system it is possible to observe spring-like mechanical oscillations as the optically bound equilibrium is restored after displacement.

It is straightforward to show, through a Taylor series expansion, that a very good leading-order approximation to the pair potential energy function, close to a local minimum, emerges in the second derivative of the above equations (12.1) and (12.2). Accordingly, treating the associated vibrations as having a simple harmonic form enables effective force constants to be calculated for the bound pair [22]. For example, in a typical study of an optically bound lattice of 1 μm silica spheres, formed in aqueous suspension at the focus of 1064 nm Nd:YAG laser radiation [23], the effective force constants were calculated to have values of around 2 pN μm^{-1}. As shown by the above expressions, the magnitude depends on not only the local geometry but also the local irradiance; in these experiments the estimated focus intensity is ~ 50 kW cm^{-2}.

12.4 Potential energy landscapes

Most of the theory on optical binding concerns laser-induced interactions between spherical particles. For progressively more intricate systems, the complexity of the theory quickly escalates, compounded by additional difficulty when three or more particles experience collective optical binding. For example, even a pair of cylindrical particles has seven degrees of spatial freedom; the associated analysis, based on carbon nanotubes, was first tackled by the present authors [24]. For nanotubes of 200 nm length and 0.4 nm radius, separated by a distance of 2 nm and irradiated by a light beam with 1×10^{12} W cm^{-2} intensity, it was determined that the

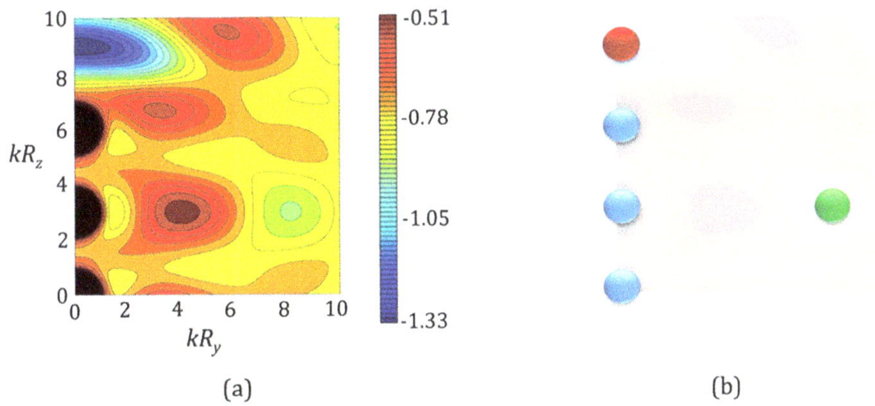

Figure 12.5. (a) Potential energy landscape for pairwise optical binding of four identical particles in the *y*, *z*-plane. Polarisation and wave-vector of the input beam are along the *x*- and *y*-axis, respectively. Three particles are already in positions fixed by optical binding potentials, as represented by the black shapes at $(kR_y, kR_z) = (0, 0), (0, 3), (0, 6)$. The fourth particle is drawn to a minimum on the resulting potential energy surface; as a result, it most likely rests at (0, 9) or (8, 3). Colour scale (arbitrary units) on the right. (b) Two stable cluster arrangements based on the potential energy landscape of (a). The three fixed particles are represented by blue circles. The red circle signifies the fourth particle in its most favourable position (the absolute energy minimum); the green circle indicates the same particle in another stable location (a local minimum). Image (a) is reprinted with permission from reference [26], copyright 2008, by the American Physical Society.

optical binding force range is 10^{-5}–10^{-12} N (according to geometry). Analogous to trapping, the optical binding of cylindrical nanoparticles is experimentally more readily achievable because one of its dimensions extends into the micron size range. Nevertheless, for the common case of spherical particles, a considerable amount of groundwork has been accomplished—leading to the mapping of potential energy landscapes [25, 26], such as those illustrated in figure 12.5. Within such energy surfaces the particles will reside at the minima, in positions where the optical binding force is null, so that certain stable arrangements can be maintained with fixed particle–particle separations.

This capacity for particles to be held together by light, in stable and non-contact geometries, bears a superficial similarity to the way in which atoms are bound together by Coulombic forces, in molecules of distinct geometry. Recognition of this analogy has led to some, perhaps over-enthusiastic, proclamations of a new form of 'optical matter' [27]. However, this exaggeration does not detract from genuine possibilities for optical self-assembly, driven by the forces of optical binding. In various experiments, the formation of two- and three-dimensional metal clusters has been observed [28–33], whose nanoparticle arrangements align with the predictions of optical landscape theory [26, 28]. In others, self-assembled sub-micron polystyrene spheres have been created with square [34] or hexagonal [35] symmetry. At present, the largest optical matter assembly comprises up to 60 nanoparticles; this is depicted in figure 12.6. Formation of such lattices is a consequence of the ensemble

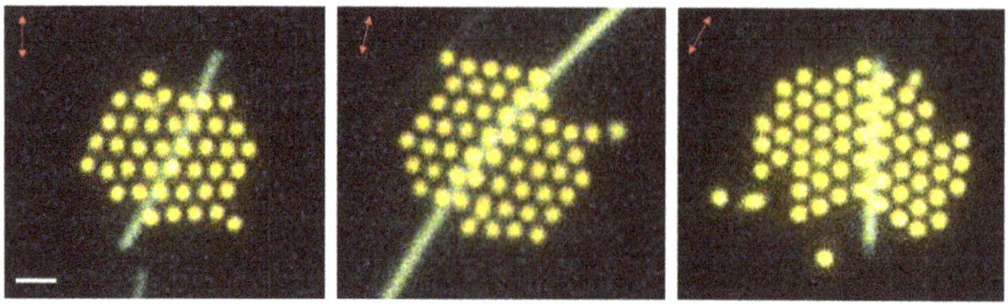

Figure 12.6. Silver-nanowire-assisted optical assembly of multiple gold nanoparticles. The stable and large-scale non-contact assembly of nanoparticles (yellow circles) forms due to the presence of the nanowire (green line), which acts as a plasmonic antenna. The polarization direction (the red arrows) of the incident light is set at a 30° angle relative to the long axis of the nanowire. The scale bar (white line) denotes a length of 1 μm. Reprinted with permission of Wiley, from reference [33], copyright 2019.

(a) (b)

Figure 12.7. Depictions of the optical matter machine. (a) Seven-nanoparticle optical matter array acting as a gear, converting spin angular momentum into orbital angular momentum that is imparted to a probe particle (outside of the incident beam). (b) Streamlines of the transverse Poynting vector in and around the gear unit. Reprinted with permission from reference [36], copyright The Optical Society.

adopting a minimum of net potential energy; the particles are naturally separated into non-contact positions that are distinct from the close-packing exhibited in many familiar physical structures, such as the optical lattice structures discussed in chapter 8 (for which the inter-particle interactions are minimal).

An interesting recent application of optical matter has been provided by Scherer and co-workers, who cleverly exploit a process in which spin angular momentum is converted into orbital angular momentum [36]. This optical matter machine works on the principle that input circularly polarised light both creates the optical matter and rotates the whole rigid body in the opposite direction to the rotation of the polarisation (figure 12.7); this process is known as 'negative optical torque' [37].

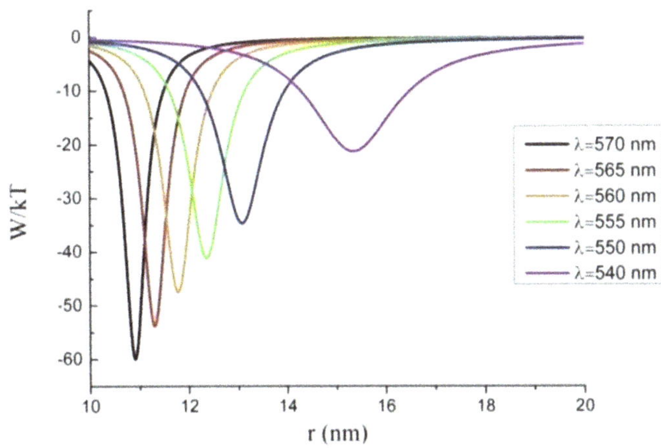

Figure 12.8. Potential energy distributions for the pairwise optical binding between 2.9 μm diameter cadmium telluride quantum dots, exhibiting substantial variation of the potential energy W (in units of k_BT) within a small range of laser wavelengths λ. The shifting positions of the energy minima dictate the wavelength-dependence of the pair equilibrium separation. Reprinted with permission from reference [50], copyright The Optical Society.

In practice, the optical binding of nanomaterials is, at present, typically restricted to metal nanoparticles [38–40]—since their larger polarisabilities, compared to dielectric substances, enable the production of a binding force that is more amenable to experimental measure. Moreover, as with optical tweezing, most experiments involve particles immersed in a liquid medium [41–44] (although, in connection with microdroplets, optical binding in air has also been reported [45, 46]). In general, any beam (or beams) that are involved in producing optical binding effects will usually have first acted as traps for the particles individually. It is therefore necessary to also account for the fact that, alongside the gradient force which moves the particles within useful range of each other, there is also a radiation force tending to push each of them along the beam axis. Such potential problems with radiation forces, in optical binding experiments, can be solved by using counter-propagating beams [47, 48]; in a set-up remarkably similar to the first optical trap created by Ashkin in 1970 [49].

One other feature of optical binding merits closer attention: its wavelength-dependence. The explicit dependence of the potential energy functions on $k = 2\pi/\lambda$, shown in equations (12.1) and (12.2) and evident in figures 12.4 and 12.5, correctly indicates that tuning the laser wavelength will shift the positions of stable equilibrium—but that is not the whole story. The dispersion (wavelength-dependence) behaviour of quantum dots is itself directly linked to particle size, and it thereby influences the magnitude of the dynamic polarisability, which features in the optical binding equations. An especially interesting aspect arises in quantum dot systems, as for example in a study of CdTe quantum dots of diameter 2.9 nm, with incident laser wavelengths in the range 550–565 nm [50]. The results (some of which are shown in figure 12.8) indicate scope for novel sorting applications.

References

[1] Forbes K A, Bradshaw D S and Andrews D L 2020 Optical binding of nanoparticles *Nanophotonics* **9** 1–17

[2] Eisenschitz R and London F 1930 Über das verhältnis der van der waalsschen kräfte zu den homöopolaren bindungskräften *Z. Phys.* **60** 491–527

[3] London F 1930 Zur theorie *und* systematik der molekularkräfte *Z. Phys.* **63** 245–79

[4] London F 1937 The general theory of molecular forces *J. Chem. Soc. Faraday Trans.* **33** 8–26

[5] Hamaker H C 1937 The London–van der Waals attraction between spherical particles *Physica* **4** 1058–72

[6] Casimir H B G and Polder D 1948 The influence of retardation on the London–van der Waals forces *Phys. Rev.* **73** 360–72

[7] Andrews D L and Bradshaw D S 2014 The role of virtual photons in nanoscalc photonics *Ann. Phys. (Berlin)* **526** 173–86

[8] Burns M M, Fournier J-M and Golovchenko J A 1989 Optical binding *Phys. Rev. Lett.* **63** 1233–6

[9] Thirunamachandran T 1980 Intermolecular interactions in the presence of an intense radiation field *Mol. Phys.* **40** 393–9

[10] Bradshaw D S and Andrews D L 2005 Optically induced forces and torques: interactions between nanoparticles in a laser beam *Phys. Rev. A* **72** 033816

[11] Bradshaw D S and Andrews D L 2005 Interactions between spherical nanoparticles optically trapped in Laguerre–Gaussian modes *Opt. Lett.* **30** 3039–41

[12] Karásek V and Zemánek P 2007 Analytical description of longitudinal optical binding of two spherical nanoparticles *J. Opt. A: Pure Appl. Opt.* **9** S215–20

[13] Karásek V, Čižmár T, Brzobohatý O, Zemánek P, Garcés-Chávez V and Dholakia K 2008 Long-range one-dimensional longitudinal optical binding *Phys. Rev. Lett.* **101** 143601

[14] Rodríguez J and Andrews D L 2008 Optical binding and the influence of beam structure *Opt. Lett.* **33** 2464–6

[15] Rodríguez J and Andrews D L 2009 Influence of the state of light on the optically induced interparticle interaction *Phys. Rev. A* **79** 022106

[16] Figliozzi P, Sule N, Yan Z, Bao Y, Burov S, Gray S K, Rice S A, Vaikuntanathan S and Scherer N F 2017 Driven optical matter: dynamics of electrodynamically coupled nanoparticles in an optical ring vortex *Phys. Rev. E* **95** 022604

[17] Figliozzi P, Peterson C W, Rice S A and Scherer N F 2018 Direct visualization of barrier crossing dynamics in a driven optical matter system *ACS Nano* **12** 5168–75

[18] McCormack P, Han F and Yan Z 2018 Self-organization of metal nanoparticles in light: electrodynamics–molecular dynamics simulations and optical binding experiments *J. Phys. Chem. Lett.* **9** 545–9

[19] Jani R, Das S C, Zahura F, Islam H, Al-Quaderi G D and Mahdy M R C 2021 Plasmonic octamer objects: reversal of near-field optical binding force without the aid of backgrounds *Appl. Opt.* **60** 10124–31

[20] Rodríguez J and Andrews D L 2009 Manipulation of optically fabricated particle arrays using broadband radiation *Proc. SPIE* **7400** 740011

[21] Smith S N A, Coles M M and Andrews D L 2011 Optical binding with anisotropic particles: resolving the forces and torques *Proc. SPIE* **8097** 80971E

[22] Metzger N, Marchington R, Mazilu M, Smith R, Dholakia K and Wright E 2007 Measurement of the restoring forces acting on two optically bound particles from normal mode correlations *Phys. Rev. Lett.* **98** 068102
[23] Han X, Luo H, Xiao G and Jones P H 2016 Optically bound colloidal lattices in evanescent optical fields *Opt. Lett.* **41** 4935–8
[24] Andrews D L and Bradshaw D S 2005 Laser-induced forces between carbon nanotubes *Opt. Lett.* **30** 783–5
[25] Rodríguez J, Dávila Romero L C and Andrews D L 2007 Optically induced potential energy landscapes *J. Nanophoton.* **1** 019503
[26] Rodríguez J, Dávila Romero L C and Andrews D L 2008 Optical binding in nanoparticle assembly: potential energy landscapes *Phys. Rev. A* **78** 043805
[27] Burns M M, Fournier J-M and Golovchenko J A 1990 Optical matter: crystallization and binding in intense optical fields *Science* **249** 749–54
[28] Yan Z, Gray S K and Scherer N F 2014 Potential energy surfaces and reaction pathways for light-mediated self-organization of metal nanoparticle clusters *Nat. Commun.* **5** 3751
[29] Huang N, Martínez L J, Jaquay E, Nakano A and Povinelli M L 2015 Optical epitaxial growth of gold nanoparticle arrays *Nano Lett.* **15** 5841–5
[30] Coursault D, Sule N, Parker J, Bao Y and Scherer N F 2018 Dynamics of the optically directed assembly and disassembly of gold nanoplatelet arrays *Nano Lett.* **18** 3391–9
[31] Kudo T, Yang S-J and Masuhara H 2018 A single large assembly with dynamically fluctuating swarms of gold nanoparticles formed by trapping laser *Nano Lett.* **18** 5846–53
[32] Nan F and Yan Z 2019 Tuning nanoparticle electrodynamics by an optical-matter-based laser beam shaper *Nano Lett.* **19** 3353–8
[33] Nan F and Yan Z 2019 Silver-nanowire-based interferometric optical tweezers for enhanced optical trapping and binding of nanoparticles *Adv. Funct. Mater.* **29** 1808258
[34] Jaquay E, Martínez L J, Mejia C A and Povinelli M L 2013 Light-assisted, templated self-assembly using a photonic-crystal slab *Nano Lett.* **13** 2290–4
[35] Mellor C D, Fennerty T A and Bain C D 2006 Polarization effects in optically bound particle arrays *Opt. Express* **14** 10079–88
[36] Parker J, Peterson C W, Yifat Y, Rice S A, Yan Z, Gray S K and Scherer N F 2020 Optical matter machines: angular momentum conversion by collective modes in optically bound nanoparticle arrays *Optica* **7** 1341–8
[37] Han F, Parker J A, Yifat Y, Peterson C, Gray S K, Scherer N F and Yan Z 2018 Crossover from positive to negative optical torque in mesoscale optical matter *Nat. Commun.* **9** 4897
[38] Demergis V and Florin E-L 2012 Ultrastrong optical binding of metallic nanoparticles *Nano Lett.* **12** 5756–60
[39] Yan Z, Shah R A, Chado G, Gray S K, Pelton M and Scherer N F 2013 Guiding spatial arrangements of silver nanoparticles by optical binding interactions in shaped light fields *ACS Nano* **7** 1790–802
[40] Yan Z, Manna U, Qin W, Camire A, Guyot-Sionnest P and Scherer N F 2013 Hierarchical photonic synthesis of hybrid nanoparticle assemblies *J. Phys. Chem. Lett.* **4** 2630–6
[41] Bowman R W and Padgett M J 2013 Optical trapping and binding *Rep. Prog. Phys.* **76** 026401
[42] Mohanty S K, Andrews J T and Gupta P K 2004 Optical binding between dielectric particles *Opt. Express* **12** 2746–53

[43] Dholakia K and Zemanek P 2010 Gripped by light: optical binding *Rev. Mod. Phys.* **82** 1767–91

[44] Brzobohatý O, Čižmár T, Karásek V, Šiler M, Dholakia K and Zemánek P 2010 Experimental and theoretical determination of optical binding forces *Opt. Express* **18** 25389–402

[45] Guillon M, Moine O and Stout B 2006 Longitudinal optical binding of high optical contrast microdroplets in air *Phys. Rev. Lett.* **96** 143902

[46] Guillon M and Stout B 2008 Optical trapping and binding in air: imaging and spectroscopic analysis *Phys. Rev. A* **77** 023806

[47] Tatarkova S A, Carruthers A E and Dholakia K 2002 One-dimensional optically bound arrays of microscopic particles *Phys. Rev. Lett.* **89** 283901

[48] Singer W, Frick M, Bernet S and Ritsch-Marte M 2003 Self-organized array of regularly spaced microbeads in a fiber-optical trap *J. Opt. Soc. Am. B* **20** 1568–74

[49] Ashkin A 1970 Acceleration and trapping of particles by radiation pressure *Phys. Rev. Lett.* **24** 156–9

[50] Tsipotan A S, Gerasimova M A, Slabko V V and Aleksandrovsky A S 2016 Laser-induced wavelength-controlled self-assembly of colloidal quasi-resonant quantum dots *Opt. Express* **24** 11145–50

David L Andrews and David S Bradshaw

Chapter 13

Past, present and future

Since ancient times, humankind has wrestled with conceiving the true nature of light and its interactions. Some of the earliest attempts to understand vision were based on light supposedly being involved in a tangible influence, in ways that would be hard to square with any modern scientific view. A classic example from 1st century B.C. is to be found in the all-embracing explanation of the world entitled *De Rerum Natura* by Lucetius [1]. Such views owed their origin to attempted rationalisations of how distant objects could possibly engage with the eye, without mechanical contact. Once such ideas were dispelled, through renaissance advances in science, it was found to be more than a curiosity when the notion of mechanical interactions was eventually put on a firm scientific basis, primarily through the pioneering work on theory by Bartoli, Maxwell, Poynting and others. It was no surprise, however, to recognize that on Earth the levels of force associated with sunlight—or indeed any other common source of light, were well below the threshold of everyday perception. As we observed in chapter 1, the first experiments confirming that theoretical work came in 1901, in work by Lebedev [2] as well as by Nichols and Hull [3]. However, it was with the arrival of the laser more than half a century later that practical applications in the form of optical tweezers first began to emerge [4].

Sixty years further on, the transformed field of optical forces has expanded enormously, and is even now being extended and reinvigorated in a variety of wholly unanticipated directions. A number of these reflect advances in nanotechnology and precision engineering, which now enable the manipulation of the tiniest amounts of matter with unprecedented precision. In the realm of physics, as we have seen, table-top magneto-optical devices now enable atoms to be trapped at microkelvin temperatures, expediting quantum information applications [5, 6]. In chemistry laboratories, lab-on-a-chip systems harness optical forces to achieve non-contact stirring and molecular separation [7]. In biology, the tweezing application of optical forces enable the remote manipulation and control of cellular components *in vivo* [8]. Yet it is not only these developments that have unfolded new applications. Advances in laser

doi:10.1088/978-0-7503-4191-2ch13

technology enable the circumvention of long-held tenets in conventional optics—such as the overcoming of the long-supposed impossibility of beating the diffraction limit [9, 10], the capacity to produce beams that are diffraction-free [11–13] or appear to defy the law of rectilinear propagation [14–16] and even the possibility of beams with polarisation states that vary in free space [17, 18]. Even the quantum limits of precision manipulation are now being approached: harnessing feedback from the detection of scattered light, it has been shown that silica nanospheres can be optically manipulated to almost the Heisenberg positional limit [19, 20].

The range of different mechanisms at work in optical manipulation continues to evolve, and this second edition of our book has expanded to accommodate wide-ranging illustrations of many of the latest advances. Our aim remains the same: to make clear the myriad connections to the principles and formulations of photonics theory. As we move on in this 'century of the photon', it is our hope that by providing a systematic description of past and current innovations, the same robust framework will continue to provide a basis to understand new effects that undoubtedly lie, at present, only just beyond the horizon.

References

[1] Carus T L and Stallings A 2007 *The Nature of Things* (London: Penguin Books)

[2] Lebedev P N 1901 Experimental examination of light pressure *Ann. Phys.* (Berlin) **6** 433

[3] Nichols E F and Hull G 1901 A preliminary communication on the pressure of heat and light radiation *Phys. Rev.* **13** 307–20

[4] Ashkin A 1970 Acceleration and trapping of particles by radiation pressure *Phys. Rev. Lett.* **24** 156–9

[5] Schmiedmayer J, Folman R and Calarco T 2002 Quantum information processing with neutral atoms on an atom chip *J. Mod. Opt.* **49** 1375–88

[6] Pepino R A 2021 Advances in atomtronics *Entropy* **23** 534

[7] Erickson D, Serey X, Chen Y-F and Mandal S 2011 Nanomanipulation using near field photonics *Lab Chip* **11** 995–1009

[8] Norregaard K, Jauffred L, Berg-Sørensen K and Oddershede L B 2014 Optical manipulation of single molecules in the living cell *Phys. Chem. Chem. Phys.* **16** 12614–24

[9] Hell S W and Wichmann J 1994 Breaking the diffraction resolution limit by stimulated emission: stimulated-emission-depletion fluorescence microscopy *Opt. Lett.* **19** 780–2

[10] Adams W, Sadatgol M and Güney D Ö 2016 Review of near-field optics and superlenses for sub-diffraction-limited nano-imaging *AIP Adv.* **6** 100701

[11] Bouchal Z and Olivík M 1995 Non-diffractive vector Bessel beams *J. Mod. Opt.* **42** 1555–66

[12] Khonina S N, Kazanskiy N L, Karpeev S V and Butt M A 2020 Bessel beam: significance and applications - a progressive review *Micromachines* **11** 997

[13] Grunwald R and Bock M 2020 Needle beams: a review *Adv. Phys.: X* **5** 1736950

[14] Siviloglou G A, Broky J, Dogariu A and Christodoulides D N 2007 Observation of accelerating Airy beams *Phys. Rev. Lett.* **99** 213901

[15] Zhang Y, Zhong H, Belić M R and Zhang Y 2017 Guided self-accelerating Airy beams: a mini-review *Appl. Sci.* **7** 341

[16] Efremidis N K, Chen Z, Segev M and Christodoulides D N 2019 Airy beams and accelerating waves: an overview of recent advances *Optica* **6** 686–701

[17] Schulz S A, Machula T, Karimi E and Boyd R W 2013 Integrated multi vector vortex beam generator *Opt. Express* **21** 16130–41
[18] Singh K, Buono W T, Forbes A and Dudley A 2021 Accelerating polarization structures in vectorial fields *Opt. Express* **29** 2727–37
[19] Magrini L, Rosenzweig P, Bach C, Deutschmann-Olek A, Hofer S G, Hong S, Kiesel N, Kugi A and Aspelmeyer M 2021 Real-time optimal quantum control of mechanical motion at room temperature *Nature* **595** 373–7
[20] Tebbenjohanns F, Mattana M L, Rossi M, Frimmer M and Novotny L 2021 Quantum control of a nanoparticle optically levitated in cryogenic free space *Nature* **595** 378–82

Further reading

Chapters 1–3

Andrews D L 2015 *Photonics, Fundamentals of Photonics and Physics* (New York: Wiley)

Andrews D L and Bradshaw D S 2018 *Introduction to Photon Science and Technology* (Bellingham, WA: SPIE Press)

Grynberg G, Aspect A and Fabre C 2010 *Introduction to Quantum Optics: From the Semi-Classical Approach to Quantized Light* (Cambridge: Cambridge University Press)

Novotny L and Hecht B 2012 *Principles of Nano-Optics* 2nd edn (Cambridge: Cambridge University Press)

Saleem M 2016 *Quantum Mechanics* (Bristol: Institute of Physics Publishing)

Chapter 4

Metcalf H J and van der Straten P 2012 *Laser Cooling and Trapping* (New York: Springer)

Proukakis N P, Snoke D W and Littlewood P B 2017 *Universal Themes of Bose–Einstein Condensation* (Cambridge: Cambridge University Press)

Saif F and Watanabe S 2022 *Optical Forces on Atoms* (Bristol: Institute of Physics)

Ueda M 2010 *Fundamentals and New Frontiers of Bose–Einstein Condensation* (Singapore: World Scientific)

Chapters 5 and 6

Ashkin A 2006 *Optical Trapping and Manipulation of Neutral Particles Using Lasers* (Singapore: World Scientific)

Gouesbet G and Gréhan G 2017 *Generalized Lorenz-Mie Theories* 2nd edn (Berlin: Springer)

Jones P H, Maragò O M and Volpe G 2015 *Optical Tweezers: Principles and Applications* (Cambridge: Cambridge University Press)

Padgett M J, Molloy J and McGloin D 2010 *Optical Tweezers: Methods and Applications* (Boca Raton, FL: CRC Press)

Chapter 7

Berns M and Greulich K 2007 *Laser Manipulation of Cells and Tissues* (New York: Academic)

Gennerich A 2016 *Optical Tweezers: Methods and Protocols* (Berlin: Springer)

doi:10.1088/978-0-7503-4191-2ch14

Glückstad J and Palima D 2017 *Light Robotics – Structure-mediated Nanobiophotonics* (Amsterdam: Elsevier)

Sun Y and Li X 2015 *Micro- and Nanomanipulation Tools* (Weinheim: Wiley-VCH)

Tassieri M 2016 *Microrheology with Optical Tweezers: Principles and Applications* (Singapore: Jenny Stanford)

Chapter 8

Fuentes B J 2012 *Optical Lattices: Structures, Atoms and Solitons* (Hauppauge, NY: Nova Science Publishers)

Lewenstein M, Sanpera A and Ahufinger V 2012 *Ultracold Atoms in Optical Lattices: Simulating Quantum Many-Body Systems* (Oxford: Oxford University Press)

Chapters 9 and 10

Allen L, Barnett S M and Padgett M J 2016 *Optical Angular Momentum* (Boca Raton, FL: CRC Press)

Andrews D L 2011 *Structured Light and its Applications: An Introduction to Phase-Structured Beams and Nanoscale Optical Forces* (Amsterdam: Academic)

Andrews D L and Babiker M 2012 *The Angular Momentum of Light* (Cambridge: Cambridge University Press)

Enoch S and Bonod N 2012 *Plasmonics: From Basics to Advanced Topics* (Berlin: Springer)

Maradudin A A, Sambles J R and Barnes W L 2014 *Modern Plasmonics* (Amsterdam: Elsevier)

Chapter 11

Fainman Y, Lee L, Psaltis D and Yang C 2010 *Optofluidics: Fundamentals, Devices, and Applications* (New York: McGraw-Hill Inc)

Hawkins A R and Schmidt H 2010 *Handbook of Optofluidics* (Boca Raton, FL: Taylor and Francis)

Rabus D G, Rebner K and Sada C 2018 *Optofluidics: Process Analytical Technology* (Berlin: De Gruyter)

Chapter 12

Salam A 2010 *Molecular Quantum Electrodynamics. Long-Range Intermolecular Interactions* (Hoboken, NJ: Wiley)

Taylor J M 2011 *Optical Binding Phenomena: Observations and Mechanisms* (Berlin: Springer)

Index

www.ingramcontent.com/pod-product-compliance
Ingram Content Group UK Ltd.
Pitfield, Milton Keynes, MK11 3LW, UK
UKHW051849100926
8556IPUK00005B/540

9 780750 341899